99-PUM-1

Evaluating Risks and Benefits of Soil Amendments Used in Agriculture

by:

Lynne H. Moss, P.E.
Camp Dresser & McKee Inc.

Eliot Epstein, Ph.D.
Tetra Tech

Terry Logan, Ph.D.
Logan Environmental Inc.

2002

Water Environment Research Foundation

The Water Environment Research Foundation, a not-for-profit organization, funds and manages water quality research for its subscribers through a diverse public-private partnership between municipal utilities, corporations, academia, industry, and the federal government. WERF subscribers include municipal and regional water and wastewater utilities, industrial corporations, environmental engineering firms, and others that share a commitment to cost-effective water quality solutions. WERF is dedicated to advancing science and technology addressing water quality issues as they impact water resources, the atmosphere, the lands, and quality of life.

For more information, contact:
Water Environment Research Foundation
601 Wythe Street
Alexandria, VA 22314-1994
Tel: (703) 684-2470
Fax: (703) 299-0742
www.werf.org
werf@werf.org

This report was co-published by the following organizations. For non-subscriber sales information, contact:

Water Environment Federation
601 Wythe Street
Alexandria, VA 22314-1994
Tel: (800) 666-0206
Tel: (703) 684-2452
Fax: (703) 684-2492
www.wef.org
pubs@wef.org

IWA Publishing
Alliance House, 12 Caxton Street
London SW1H 0QS, United Kingdom
Tel: +44 (0) 20 7654 5500
Fax: +44 (0) 20 7654 5555
www.iwapublishing.com
publications@iwap.co.uk

Library of Congress Catalog Card Number: 2002112606

WERF ISBN: 978-1-893664-59-3 IWAP ISBN: 978-1-84339-651-2 WEF ISBN: 978-1-57278-207-5

Camp, Dresser & McKee Inc.

ACKNOWLEDGMENTS

Report Preparation

Principal Investigator:
Lynne H. Moss, P.E.
Camp Dresser & McKee Inc.

Project Team:
Eliot Epstein, Ph.D.
Tetra Tech

Terry Logan, Ph.D.
Logan Environmental Inc.

Stephen D. Frank, A.P.R

Karen Scott
KNS Editorial Consultants

Project Subcommittee

Michael D. Jawson, Ph.D.
U.S. Department of Agriculture, Agricultural Research Service

Raymond J. Kearney, P.E.
City of Los Angeles

Barry T. Kintzer, P.E.
U.S. Department of Agriculture, Natural Resource and Conservation Service

Stephen Clough, Ph.D.

Alan B. Rubin, Ph.D.
U.S. Environmental Protection Agency

Project Reviewers

Robert Dowdy, Ph.D.
US Department of Agriculture, Agricultural Research Service (Retired)

Wayne R. Kussow, Ph.D.
University of Wisconsin at Madison

Water Environment Research Foundation Staff

Research Program Director: Jeff Moeller, P.E.
Project Manager: Lola Olabode, M.P.H.

ABSTRACT AND BENEFITS

Abstract:

Farmers today are faced with an array of products that they can use to improve the fertility of their lands. Organic soil amendments such as biosolids, manures and, to a lesser extent, municipal solid waste (MSW) composts are products that can be used to supplement or replace mineral fertilizers. Many farmers are aware of the benefits that organic products offer, especially in terms of improved crop yields, increased soil fertility, reduced fertilizer costs, and their overall contribution to sustainable agriculture.

Concerns have been raised, however, regarding the safety of organic soil amendments. Depending on the type of product, these can include (but are not limited to) the accumulation of phosphorus and metals in soils, plant uptake of land-applied metals, the contamination of groundwater and soil with organic compounds, and potential health impacts from pathogens and viruses.

Although many of these issues have been rigorously researched, data is often published in academic journals and textbooks, and is not necessarily accessible to the public. For all soil amendments, comparative information regarding risks, benefits, advantages, and disadvantages is not always readily available. Equally critical is the lack of accessible comparative data on soil amendment and mineral fertilizer characteristics and use.

Recognizing the need to provide a comparison of these products, the Water Environment Research Foundation (WERF) commissioned this multiyear study of soil amendments, fertilizers and their uses. For the study, the project team performed a comprehensive literature survey, reviewing and compiling the results from nearly 500 documents (primarily from academic journals and texts). The study generally found that the relative risk to the environment from amendments and fertilizers varies by parameter and shows that known risks from each of the materials studied can be managed. Moreover, these manageable risks must be carefully weighed against the considerable benefits provided by the land application of amendments and fertilizers.

Benefits:

- This research project offers an accessible summary of amendment and mineral fertilizer characteristics, benefits, and risks.
- The long-term field studies upon which this research is based demonstrate the relative benefits and risks of amendments and fertilizers used in agriculture.
- The report identifies research needs for soil amendments.

Keywords: Biosolids, manures, mineral fertilizers, land application, benefits and risks

TABLE OF CONTENTS

LIST OF TABLES

LIST OF FIGURES

EXECUTIVE SUMMARY

Farmers today are faced with an array of products that they can use to improve the fertility of their lands. Organic soil amendments such as biosolids, municipal solid waste (MSW) composts, and manures are products that can be used to supplement or replace mineral fertilizers. Many farmers are aware of the benefits that organic products offer, especially in terms of improved crop yields, increased soil fertility, reduced fertilizer costs, and their overall contribution to sustainable agriculture.

Concerns have been raised, however, regarding the safety of organic soil amendments. Depending on the type of product, these can include (but are not limited to) the accumulation of metals in soils, plant uptake of land-applied metals, the contamination of groundwater and soil with organic compounds, and potential health impacts from pathogens and viruses. These questions and concerns regarding amendments must be answered, and those answers effectively communicated in order to restore (or generate) confidence in the use of organic soil amendments.

Recognizing the need to provide the public — especially agricultural and other users of soil amendments — with credible, objective, accessible information, the Water Environment Research Foundation (WERF) commissioned a multi-year study of soil amendments and their uses. Highlights of the study, which focuses on biosolids, manures, MSW and yard waste compost, and mineral fertilizers, are presented here.

ES.1 Quantities

For more than 2,000 years, farmers have supplemented soils to increase crop yields. For most of that time, they applied animal manures and "night soil" to provide needed plant nutrients, and to enrich the soil. In the 1940s, the advent of mineral fertilizers helped prompt the "green revolution," dramatically increasing crop production in the United States and worldwide. Since the early 1900s, farmers have also used biosolids (sewage sludge processed to reduce pathogens and vector attraction) in the same manner in which they had previously used manures.

Currently, each of these materials – manures, fertilizers, and biosolids – is applied to the nation's agricultural landscape. Mineral fertilizers supply most the nation's crop nutrient needs. More than 50 million tons (45 million metric tons) are used annually (USDA, 1997), supplying slightly more than 20 million (18 million metric tons) tons of nutrients to the land. Biosolids, however, whose nutrient content is relatively small compared to mineral fertilizers, are applied to less than 1% of the nation's agricultural acreage (Draeger et al., 1999). An estimated 6.9 million dry tons (6.3 million dry metric tons) of biosolids was generated in 1998, about 41% of which was applied to agricultural lands (USEPA, 1999a). As shown in Figure ES-1, manure quantities generated and applied to the land are significant compared to mineral fertilizers and biosolids: 1992 estimates place manure generation at about 133 million dry tons (121 million dry metric tons) annually (USEPA, 1998), 90% of which are land-applied (Barker et al., 2000).

Figure ES-1
Amendment and Fertilizer Use

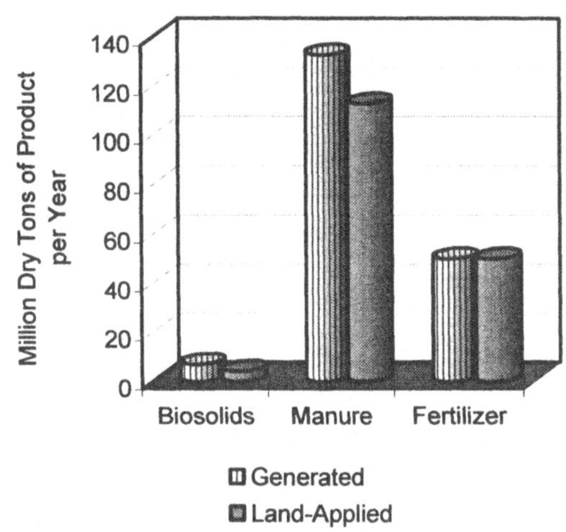

The quantity of MSW and yard waste composts used in agriculture is far less than that of the products noted above. In 1999, almost 400 million tons (363 million metric tons) of MSW was generated in the United States (Goldstein, 2000), but a 1999 BioCycle survey indicates that less than 500,000 tons (454,000 metric tons) of MSW is composted annually (Glenn and Block, 1999). ard waste composting is becoming an accepted and growing practice. The amount of material directed to agricultural applications has not been defined, however, but is likely to be low.

ES.2 Product Characteristics

The project team compiled data on physical characteristics, nutrient content and availability, metals content and relative bioavailability, and pathogen content (for organic products only). Other areas of research included the dioxin content of biosolids and mineral fertilizers, herbicides in compost, and residual antibiotics and endocrine disruptors found in amendments. Data availability varied for the products studied, but the most comprehensive databases were identified for biosolids and mineral fertilizers. Manure data were less prevalent, especially pertaining to manure metals, pathogen contents, and nutrient availability.

ES.2.1 Nutrients

The nutrient content of soil amendments is generally far lower than mineral fertilizers and, accordingly, amendments must be applied at much higher rates to meet plant nutrient requirements (or be supplemented by mineral fertilizers).

Table ES-1 provides a comparison of primary nutrient concentrations for the products assessed here.

Table ES-1
Primary Nutrient Content of Soil Amendments and Fertilizers
(%, Dry Weight Basis)

Product	Nitrogen (N)	Phosphate (P_2O_5)	Potash (K_2O)
Biosolids [(1)]			
Alkaline Stabilized (Class A)	1.0	0.9	0.4
Compost	2.8	3.9	0.4
Alkaline and Digested Liquid	5.3	5.0	0.4
Alkaline and Digested Cake	4.1	4.4	0.2
Heat Dried	6.0	7.1	0.4
Stored Manures [(2)]			
Dairy	2.1-4.8	0.7-3.6	1.4-4.8
Beef	2.0-4.8	1.3-3.6	2.2-4.8
Swine	2.5-10.8	1.9-7.5	1.4-6.6
Poultry	2.5-7.2	3.0-5.4	1.6-3.8
Mineral Fertilizers [(3)]	15-82	8-76	22-62

(1) Adapted from Lue-Hing et al. (1998)

(2) Adapted from American Society of Agricultural Engineers Standards (ASAE), 2000

(3) The Fertilizer Institute, 1982

Biosolids and manures also contain essential secondary nutrients (calcium, magnesium and sulfur) and micronutrients (boron, chlorine, copper, iron, manganese, molybdenum, and zinc) that are also beneficial to plants.

In addition to nutrient content, nutrient availability (especially nitrogen availability) is of key concern to agricultural users. Nitrogen in amendments is largely present in organic form, which must be mineralized to become available to plants. Mineralization rates can be difficult to predict, and uncertainty regarding nitrogen availability is one reason cited by farmers for not using amendments. For WERF, Gilmour et al. (2000) recently developed a comprehensive approach to help determine mineralization rates in biosolids; similar, broad scale efforts are being initiated for manures.

Reflecting the increasing focus on environmental impacts from land-applied phosphorus, research is also beginning to yield comparative information on phosphorus (P) availability in biosolids, manures, and fertilizers. In a recent and comprehensive study assessing P availability as defined by water extractable P (WEP), Brandt, Elliot, and O'Connor (2001) found that WEP in biosolids was far lower than manure or fertilizer WEP. The researchers found, for example that the WEP for triplesuperphosphate (2 samples), dairy manure (5 samples), poultry manure (8 samples) and digested biosolids (18 samples) was 85%, 48%, 21% and 3%, respectively. They attribute the lower WEP in biosolids to the removal of water soluble P during dewatering and/or the addition of iron or alum for phosphorus removal. Water treatment plant residuals, which are often discharged to the local wastewater treatment facilities, can contain high concentrations of iron or aluminum, and can immobilize P as well (Peters and Basta, 1996).

ES.2.2 Metals

Metal content was an area of particular interest, and the project team found that metal concentrations in some manures were commensurate to or greater than biosolid metal concentrations (Sims, 1995). Metals provided directly and indirectly through feed supplements are responsible in part for manure metal content. Primary nutrient mineral fertilizers generally have low metal contents, with the exception of some phosphate fertilizers, which can contain elevated concentrations of arsenic, cadmium, chromium and nickel. Micronutrient mixes, by design, also exhibit high metal contents. Table ES-2 presents metal concentrations from biosolids and some manures, while Table ES-3 shows metal concentrations in mineral fertilizers. When comparing metal contributions from land-applied materials, however, it is important to remember that application rates vary — mineral fertilizer application rates are generally much lower than biosolids or manure application rates, and the impacts from fertilizer contributions on a per site basis should be considered accordingly. The higher metal concentrations in fertilizers are generally offset by the small amount of these concentrated materials that are applied. Manure application rates are similar to application rates for biosolids, however, and therefore the applied metals can be similar for the two products on a per-site basis.

Table ES-2
Soil Amendment Metal Concentrations (mg/dry kg)

Metal	Dairy Cattle Manure (1)	Beef Cattle Manure(1)	Swine Manure(2)	Poultry Manure(3)	Biosolids Mean(4)
Arsenic	NA	NA	3.7	13	5.0
Cadmium	0.25	NA	2.5	2.4	4.4
Copper	38	36	109-501	465	425
Lead	NA	NA	7.6	46	76
Mercury	NA	NA	ND	NA	1.8
Molybdenum	6.2	4.94	2.6	19	12
Nickel	23	NA	29	16	33
Zinc	150	129	455-656	602	735

ND = not detected; NA = not available

(1) Adapted from ASAE, 2000

(2) Compiled from ASAE, 2000 and Epstein, 1999

(3) Compiled from ASAE, 2000 and Texas Agricultural Extension Service (2001)

(4) Average from data in Lue-Hing et al. (1998), weighted by number of samples

Table ES-3
Mean Heavy Metal Concentrations in Mineral Fertilizers (mg/kg)

Product	Arsenic	Cadmium	Chromium	Copper	Lead	Mercury	Nickel	Zinc
Phosphate Fertilizer	11.3	65	173.3	56.6	12.2	0.1	27.5	240.2
NPK for Phosphorus	12.8	30.6	83.9	31.4	217	0.07	28.5	234
NPK for Nitrogen	5.3	5.0	50.6	41.3	31.8	0.2	11.8	205
Potash	0.3	NA	0.5	1.0	NA	0.1	1.4	1.6
Sulfur (nutrient)	5.0	36.6	37.9	24.3	5.4	0.1	48.9	385
Sulfur (pH adjustment)	7.8	NA	NA	40.4	6.4	NA	NA	NA
Lime	14.6	2.5	21.2	42.0	44.5	0.1	7.9	353
Gypsum	4.5	1.7	1.4	23.7	4.6	0.011	3.0	53.8
Iron	1,663	177	NA	667	7,135	NA	NA	NA
Boron	531	0.75	1.3	8.1	5.5	0.2	2.5	6.0
Manganese	7.8	2.3	10.0	11.3	27.5	0.01	50	60.8
Zinc	20	398	339	1,101	10,013	3.36	890	NA
Micronutrient Mixes	41.9	27.9	230	29,650	1,798	0.1	12.5	77,300

NA = not available
From: USEPA, 1999b

It should also be noted that the metal concentrations in biosolids are regulated (by the 40 CFR 503 *Standards for the Use or Disposal of Sewage Sludge*), while metals in manures and, for the most part, fertilizers are not.

ES.2.3 Pathogens

A key difference in amendment pathogens is regulation-based. Biosolids indicator pathogen concentrations are regulated, while no such regulations exist for manures. The federal 40 CFR 503 regulations require that biosolids be treated to reduce pathogens and define two different classes of biosolids based on indicator pathogen content. (The quantification of pathogens is an evolving field and, to date, many have relied on bacteria such as fecal coliform and *Salmonella* as "indicators" of total pathogen content.) Class A products are limited to fecal coliform concentrations of no more than 1,000 most probable number per gram (MPN/g) dry solids or 3 MPN/4g dry solids of *Salmonella*. Class B products, which have use restrictions, are limited to 2,000,000 MPN/g (or colony forming units/gram, CFU/g).

The limited data available show that indicator pathogen contents of manures can exceed the pathogen content of biosolids. For example, fecal coliform concentrations in fresh manures can range from 0.5 x 10^7 to more than 3 x 10^{10} colonies/dry g in dairy, beef, swine, and some poultry manures (ASAE, 2000). Jones (1980) reported that *Salmonella* in English cattle generally are on the order of 100 cells/ gram, although 3.1×10^3 cells/ gram feces were found in infected dairy cows (Jones et al. 1982). Reductions in these levels might be expected due to the elapsed time between excretion and land application, as well as various levels of treatment (i.e, anaerobic lagoons, storage/stacking) that are employed at most livestock operations (Kintzer, 2002). Nonetheless, fecal coliforms and *Salmonella* concentrations in biosolids, especially Class A products, are lower, as shown in Table ES-4.

Table ES-4
Biosolids Pathogen Contents

Pathogen	***Class A Alkaline Stabilized***	***Class A Composted***	***Class A Heat-dried***	***40 CFR 503 Class A Standards***
No. of Wastewater Treatment Facilites in Sample	5	4	5	NA
Fecal Coliform (MPN/g Mean)	3	76	6	<1,000
Fecal Coliform (MPN/g Median)	1	506	8	NA
Salmonella sp (MPN/4g Mean)	2	2	0	<3
Salmonella sp (MPN/4g Median)	2	2	0	NA

Pathogen	***Class B Alkaline Stabilized***	***Class B Digested Liquid***	***Class B Digested Cake***	***40 CFR 503 Class B Standards***
No. of Wastewater Treatment Facilites in Sample	55	26	41	NA
Fecal Coliform (MPN/g Mean)	60	104,600	6,521	<2,000,000
Fecal Coliform (MPN/g Median)	9600	472,600	16,071	NA
Salmonella sp (MPN/4g Mean)	2070	NA	NA	NA
Salmonella sp (MPN/4g Median)	4000	NA	NA	NA

Source: Lue-Hing et al., 1998
NA = not applicable

There is very little information on pathogens in MSW and yard waste compost. For MSW compost, Golueke, Diaz, and Gurkewitz (1989) reported fecal and total coliforms reaching concentrations greater than 1.1 x 10^8 and 4 x 10^8, respectively. They did not detect *Salmonella.* With respect to yard waste compost, the authors report that "pathogens of most concern either were present in low concentrations or could not be detected." Meckes et al. (1995) found yard waste compost fecal coliforms ranging from <2.5 most probable number per gram (MPN/g) to 2, 300 MPN/g, with an average of 38 MPN/g.

ES.2.4 Other Parameters

Another chemical of special interest is dioxin, an ubiquitous compound that has been found in biosolids and fertilizers (no studies of manure dioxin concentrations have been identified). Some dioxin forms are more toxic than others, and, thus, levels of dioxin forms (or cogeners) are usually referred to by their toxic equivalent (TEQ). (The total dioxin TEQ is essentially a sum of the concentration of each dioxin form multiplied by its individual toxic equivalent factor, or TEF.) A survey by the Association of Metropolitan Sewerage Agencies (AMSA) found biosolids dioxin concentrations to be quite low, averaging 48.5 ppt-TEQ (Alvarado et al., 2001), well below the 300 ppt TEQ limit in proposed revisions to the federal 40 CFR 503 regulations. A survey of 50 fertilizers also showed low or undetectable concentrations of dioxins in most products, but several micronutrient fertilizers did have elevated dioxin concentrations (WSDE, 1999). These products, both derived from steel mill flue dust had TEQs of 140 ppt and 340 ppt.

Antibiotic residues in manures (Karpati and Rubin, 1998; Campagnolo and Rubin, 1998) and biosolids and endocrine disruptors in both biosolids and manures (Dentel et al. 1993; Shore, Correll, and Chakaroborty, 1995) have been raised as a potential concern, but the direct impact of these chemicals in land-applied amendments is not clearly understood.

Two herbicides, clopyralid and picloram, have been found in yard waste composts (Bezdicek et al., 2001). Composters are working to eliminate these chemicals, which are inadvertently introduced through feedstocks, from their composted yard wastes.

ES. 3 Benefits

The team's assessment of product benefits focused on soil amendments (biosolids, manures and composts), as the benefits of mineral fertilizers (which are manufactured specifically to meet agricultural needs) are fairly well understood. Biosolids and manures, on the other hand, are often thought of as byproducts or waste products. Since their nutrient contents are relatively low compared to mineral fertilizers, their crucial role in agriculture is less clear.

Project research, however, identified numerous studies that reinforced and emphasized the importance of soil amendment use in agriculture, most particularly its role in improving soil properties. With their high organic contents, soil amendments work to enhance soil quality and productivity in a variety of ways. Specifically, the use of soil amendments can improve soil structure, reduce erosion potential, reduce bulk density (facilitating tillage and seedling emergence), increase water retention, and increase soil cation exchange capacity (improving the soil's ability to retain nutrients) (Oberle and Keeney, 1994; Lindsay and Logan, 1998; Rynk, 1992; Gouin, 1997; Tester, 1990; Council for Agriculture Science and Technology [CAST],

1996). Improved soil properties can translate into a reduced reliance on pesticides and inorganic nutrients, which in turn can reduce potential water quality impacts from agriculture. Increased crop yields (compared to chemically fertilized fields) have also been noted for both biosolids and manures (Jacobs et al., 1993; Peterson et al., 1994; CAST, 1996).

Other notable benefits of soil amendments are their role in carbon sequestration (and the reduction of greenhouse gases), and for composts especially, plant disease suppression (Humpert, 2000; Hoitink, et al., 1997). Additionally, the micronutrients in biosolids and manures can correct deficiencies in soils (Mathers et al., 1980; McCaslin et al., 1987; Scharp, 2000).

Some benefits have not been adequately explored, however. It appears that economic evaluations of soil amendments have focused only on their "fertilizer replacement value," generally ignoring the impact of improved soil health on farm economics. Further research in this area is recommended to quantify the overall benefit of soil amendments to the farmer.

ES.4 Risks

Research indicates that there are *potential* environmental concerns associated with the use of biosolids, manures, and fertilizers. The range of environmental and health issues identified cover the entire "scale of risk" from perceived risk (for example, an issue that is a concern, but has not been demonstrated to be problem after years of practice) to large scale, demonstrated impacts. Here we summarize the concerns identified, but, critically, assess them within the framework defined by the following two questions:

- What is the magnitude of this issue?
- Can risks be managed?

ES.4.1 Nitrogen

With respect to nitrogen, fertilizers have been identified as a primary agricultural material responsible for nitrate contamination of groundwater (Bouchard, Williams and Surampalli, 1992), most especially in the midwestern United States. These authors cite over-application as the primary basis for contamination and further identify two reasons for this practice: (1) a lack of information on the farmer's part to accurately predict nitrogen needs and (2) "the economic cost of overfertilization is less than the economic cost of underfertilization." Nitrogen has also been identified as a substantial pollutant of surface waters, especially salt waters. Excess nitrogen from fertilizers, manures and other sources, for example, is a suspected contributor to a hypoxic zone (an area of depleted oxygen) in the Gulf of Mexico during the summer months where the Mississippi empties into the Gulf (Goolsbey and Battaglin, 1997).

Despite the large volume of manure applied to the land, mineral fertilizers remain the largest source of land-applied nitrogen, contributing more than 11 million tons (10 million metric tons) annually (1995 value, from USDA, 1997). Manures are estimated to add less than 6 million tons (5.4 million metric tons) of nitrogen annually (USEPA, 1998), with volatilization and denitrification further reducing this value. Less than 3 million dry tons (2.7 million dry metric tons) of biosolids is land applied (USEPA, 1999a), and with nitrogen concentrations averaging less than 5%, the relative impact of these materials is minimal.

The risks of excess nitrogen application — for all materials — can be managed through the use of nutrient management plans (NMPs), a growing practice applied to each of the materials discussed here. NMPs account for soil conditions, cropping practices, and nutrient sources to better predict nitrogen application requirements that will optimize crop yields without adversely impacting the environment.

ES.4.2 Phosphorus

Phosphorus has been implicated as a leading contributor to surface water pollution, most especially eutrophication (Sharpley et al., 1999) of freshwater. Again, each of the materials studied here contribute phosphorus to the environment. Estimated phosphate fertilizer use was 4.4 million tons (4.0 million metric tons) in the US in 1995 (USDA, 1997), while phosphate in applied manures is estimated to be about 3.6 million tons (33 million metric tons). As with nitrogen, the total contribution of biosolids-supplied P is negligible compared to fertilizer or manure sources.

With respect to environmental impacts on a per site basis, studies generally support the findings (noted earlier) that P availability is lower in biosolids than other products. A two-year British study compared phosphorus in runoff from fields amended with liquid digested biosolids, dewatered biosolids, liquid cattle manure, and the mineral fertilizer triplesuperphosphate (Withers, Clay, and Breeze, 2001). These materials were applied at maximum recommended P rates to a silty clay loam soil. The researchers found that fields receiving the triplesuperphosphate exhibited the highest P runoff, followed by cattle manure, liquid biosolids, and dewatered biosolids. In an earlier study, McLeod and Hegg (1984) also found that runoff from pastures amended with biosolids had lower P concentrations than runoff from manured or mineral fertilized fields.

Recognizing the need to manage agricultural P inputs, agricultural institutions are beginning to incorporate phosphorus management on a broad scale. For example, the Natural Resources Conservation Service (NRCS) recently revised its nutrient management standards to incorporate P management tools (such as the phosphorus index) (NRCS, 1999). The NRCS revisions address concerns regarding manure applications in particular, but are appropriate for any phosphorus (or other nutrient) source applied to the land.

ES.4.3 Metals

The fate of biosolids-applied metals remains a concern for some researchers. Specific concerns include groundwater impacts (Camobreco et al., 1996), and yield reductions (Bouldin, 1997). Long-term research for biosolids applied at typical application rates does not support these concerns, however. Many researchers have shown that biosolid-applied metals accumulate in surface soils and do not migrate to groundwater (Jacobs et al., 1993; Dowdy et al, 1994;USEPA, 1977, Garcia and Mangaroo, 1994; Barbarick, Ippolito, and Westfall, 1998). For example, Sloan, Dowdy and Dolan (1998) showed that 16 years after application, greater than 90% of biosolids-applied cadmium, copper, lead, nickel, and zinc remained in the upper 0.30 m (1 ft) of the soil, whereas 100% could be accounted for in the upper 0.45 m (1.5 ft). Comparing the migration of metals from biosolids and fertilizer applications, researchers in Colorado found that, for some metals (copper, for example), metal concentrations below surface soils (at 20–60 cm or 0.65–2.0 ft deep) from fertilizer applications were similar to metals from biosolids applications, despite the fact that the quantity of biosolids-supplied metals was greater (Barbarik,

Ippolito, and Westfall, 1998). Working with a heat-dried biosolids in Texas, researchers found that cadmium, lead and nickel concentrations in the soil (at 30–45 cm, or 1.0–1.5 ft. depth) were greater with annual applications of commercial macronutrient fertilizer than with biosolid applications (Garcia and Mangaroo, 1994), although surface soil (0–15 cm or 0–6 in. depth) concentrations were greater where biosolids had been applied.

Based upon data from a heavily loaded biosolids application site, Bouldin (1997) notes concerns regarding yield reductions in corn forage. Yet years of long-term studies in the United States demonstrate that biosolids applied at agronomic rates do not impair crop yields, but can instead produce yields that are equivalent to (or better than) yields from mineral-fertilized or manured fields. Evaluating 10 long-term biosolids application sites across the nation, researchers saw no phytotoxic effects from metals loadings, even at a site that received biosolids applications exceeding 503 Cumulative Pollutant Loading Rates for six metals (Jacobs et al., 1993). (The researchers note that models used for 503 development would predict this outcome.) Other long-term studies evaluating digested biosolids (Clapp et al., 1994), alkaline-stabilized biosolids (Hormann et al., 1994), and heat-dried biosolids (Garcia and Mangaroo, 1994) found yields similar to mineral-fertilized plots, although soil metal concentrations and plant uptake of some metals were greater in biosolids-amended fields.

With respect to animal health and land-applied metals, a key area of study appears to be the impact of biosolids on molybdenosis in ruminants. In a 3-year WERF study, O'Connor and McDowell (1999) found that the transfer of metals to forage was minimal for a variety of application rates (the minimum rate studied was the agronomic rate), and that recommended metal concentrations in feed were not exceeded. Molybdenum, a focus of the study, never exceeded 3/mg/kg and the copper: molybdenum ratio remained within acceptable levels. Concerns regarding molybdenosis (molybdenum-induced copper deficiency) were a driving force behind the study, and the cattle in the study did exhibit symptoms of copper deficiency. The high sulfur content of the forage grown on biosolids-amended soils, however, was cited as the cause of the copper deficiency, and copper supplements (given in the third year of the study) corrected the problem.

Although U.S. concerns focus on biosolids, data from Europe and Australia indicated that both fertilizers and manures are significant sources of metals, especially cadmium, in agricultural soils. For example, Van der Voet et al. (1994) found that 50% of soil cadmium in the European community was contributed by fertilizers, 3% by biosolids, 14% from atmospheric deposition and 18% from manure applications. Like biosolids-applied metals, metals from fertilizers and manures appear to accumulate in surface soils (Jindasa et al., 1997; Chang, Sommerfeldt, and Entz, 1991; Mulla, Page, and Ganje, 1980), where the potential for plant uptake is greatest. Cadmium uptake has been noted to be a specific concern in Australia, where vegetable cadmium concentrations in excess of allowable limits are attributed primarily to poultry manure applications, with phosphate fertilizers also playing a significant role (Jindasa et al., 1997). Reviewing data from the Rothamsted Experiment Station in England (perhaps the world's oldest agricultural study site, with more than 140 years' worth of data), Jones, Symon, and Johnston (1987) also note that manure is likely a greater contributor to soil cadmium than inorganic fertilizers.

The relative impact of metals applied through amendments and fertilizers in the United States has not been clearly established but can be roughly estimated by multiplying typical metal concentrations in these products and their annual application rates (Table ES-5). As indicated by the table, it appears that far more metals are applied to U.S. lands through manure applications than through biosolids or fertilizers (an exception is cadmium supplied in phosphates). We caution that, as noted earlier, the different application rates — per site — for the products listed must be considered as well when considering metal impacts. For example, although phosphates contribute more cadmium to the soil than the other products overall, that contribution is generally spread across a larger application area than biosolids or manures, and so per site contributions (though still greater than biosolids or manures) are smaller than indicated by Table ES-5. Comparisons between manures and biosolids are somewhat more valid, however, because the per-site application rates for these products are similar.

Table ES-5

Estimated Total Metals Applied to Land from Various Products

	Applied Metals (tons/year) (1)			
Metal	Biosolids (2)	Swine Manure (3)	Poultry Manure (4)	Phosphate Fertilizer (5)
Arsenic	14.2	33.3	189	49.7
Cadmium	12.6	22.5	33	286
Chromium	168	NA	NA	761
Copper	1,202	2,745	6,743	249
Lead	215	68.4	667	53.7
Mercury	5.07	NA	NA	0.0
Molybdenum	34.2	NA	NA	NA
Nickel	94.1	264	NA	121
Selenium	10.9	NA	NA	NA
Zinc	2,081	4,995	8,729	1,057

Notes:

(1) Based on metal concentrations presented in Table ES-2.

(2) Assumed application of 2.83 million dry tons annually (USEPA, 1999a)

(3) Assumed application of 9 million dry tons annually (USEPA, 1998)

(4) Assumed application of 14.5 million dry tons annually (EPA, 1998)

(5) Assumed application of 4.4 million tons annually (USDA, 1997)

Regardless, risk assessments accepted by the mainstream of the scientific community have demonstrated that metals supplied through most fertilizers and biosolids do not pose a risk to human health when applied at recommended rates. A risk assessment addressing manure metals has not yet been performed, but based upon the information presented here, would be of interest. Although it is impossible to say what the outcome of such an assessment might be, it is critical to note that, like biosolids, manures have generally been applied to agricultural soils for many years without any observed impacts on human or plant health from metals that they contain (exceptions being the Australian applications noted above and zinc phytotoxicity from the over-application of swine manure in North Carolina).

ES.4.4 Pathogens

As noted earlier, indicator pathogen concentrations are higher in manures than in biosolids and, unlike biosolids, their concentrations are not strictly regulated. The relative contribution of pathogens from biosolids and manures is an important consideration in determining environmental impact. In England, where pathogen reduction requirements for biosolids are not as strict as the United States, manures nonetheless contribute considerably more *Salmonella* to the environment than biosolids. As is also true in the United States, the larger volume of manures applied, coupled with the propensity for herd or flock infection, are cited as the basis for higher pathogen contribution of manures (WaterUK, 2000).

With respect to pathogen contamination, however, other sources must be considered as well. Wild animals and birds contribute to ambient pathogen levels through their manures and even infected carcasses (WaterUK, 2000). For example, one study found that, near a well-operated livestock operation, manures from about one-third of the wild animals trapped contained *Cryptosporidium* oocysts (Drury and Lloyd, 1997).

Simple management practices (for example, setbacks from waterways) might warrant consideration to reduce any potential risks associated with manure pathogens and, in fact, some agencies have already initiated such steps. For example, the state of Florida has a 120-day restriction on the use of consumer crops grown on manure-amended soils (Epstein, 1999).

ES.4.5 Other Chemicals

Primary concerns at this time include dioxin, which is slated to be regulated in biosolids under proposed revisions in 40 CFR 503 and endocrine disruptors (compounds that can affect human and animal reproductive systems), which are contained in biosolids and manures. Like the 503 metal limits, the proposed dioxin limit was developed through a risk assessment and, based upon dioxin concentrations reported to date (Landers, 2000), it is expected that the vast majority of biosolids generated will be able to meet the proposed limit. Dioxin limits have not been established for fertilizers, although (as noted earlier) high dioxin concentrations have been found in some micronutrients made from recycled hazardous wastes; these products do not represent the bulk of fertilizers generated, however.

Little is currently known about direct impacts of biosolids or manure-supplied endocrine disruptors on the environment and much further study is required in this area. This issue is complex, however, with multiple stressors potentially acting in concert to produce a given impact (NRC, 2000). Trying to determine the specific effect of a single compound, or even several related compounds, is not a simple undertaking and extensive research may be required to address this issue.

ES.4.6 Odors and Other Emissions

Odors are an acknowledged part of biosolids and manure land applications. To date, odor has primarily been viewed as a nuisance issue, but increasingly, the health impacts of odors from land application sites, as well as other application emissions (such as bacteria) are now being questioned, despite the fact that no documented health impacts have been identified. Although at least one study has shown that airborne pathogens from Class B biosolids are a potential risk for workers at biosolids application sites (Dowd et al., 2000), the authors note that the risk was small and, potentially, overestimated. No similar studies have been performed for manures.

ES.5 Summary

As noted above, the primary purpose of this study was to gather existing data on amendment benefits and risks (especially in comparison to mineral fertilizers) into a concise, readable format. This study also demonstrates, however, that there are numerous benefits associated with amendment use, and that known risks can be managed. In deciding what products to use, it is critical to balance the risks associated with the use of both amendments and fertilizers with the benefits they offer. For all products studied, proper application is the key to optimizing these benefits, while also protecting the environment as well.

CHAPTER 1.0

INTRODUCTION

1.1 Background and Purpose

For more than 2,000 years, farmers have supplemented soils to increase crop yields. For most of that time, they have applied primarily animal manures to provide needed plant nutrients, and to enrich the soil. In the 1940s, the advent of mineral fertilizers helped prompt the "green revolution," dramatically increasing crop production in the United States and worldwide. Since the turn of the century, farmers have also used *biosolids* (sewage sludge processed to reduce pathogens and vector attraction) in the same manner in which they had previously used manures.

Currently, each of these materials — manures, fertilizers, and biosolids — is applied to the nation's agricultural landscape. Mineral fertilizers supply most of the nation's crops nutrient needs. More than 50 million tons (45 million metric tons) are used annually (USDA, 1997), supplying slightly more than 20 million tons (18 million metric tons) of nutrients to the land. Biosolids, however, whose nutrient content is relatively small compared to mineral fertilizers, are applied to less than 1% of the nation's agricultural acreage (Draeger et al, 1999). An estimated 6.9 million dry tons (6.3 million dry metric tons) of biosolids were generated in 1998, about 41% of which were applied to agricultural lands (USEPA, 1999a). As Figure 1-1 shows, manure quantities generated and applied to the land are significant compared to mineral fertilizers and biosolids: 1992 estimates place manure generation at slightly more than 133 million dry tons (121 million dry metric tons) annually (USEPA, 1998), of which about 90% are land-applied (Barker, Stratton, and Rechcigl, 2000).

Fertilizers and organic soil amendments, including biosolids and manures, benefit agriculture when they are used appropriately, but they can also pose potential environmental and health risks if managed improperly. This document summarizes these benefits and risks in a format that will facilitate agricultural users' ability to determine which product or products might best suit their needs. This document also addresses specific questions that farmers and the general public have raised about organic amendments.

Figure 1-1
Amendment and Fertilizer Use

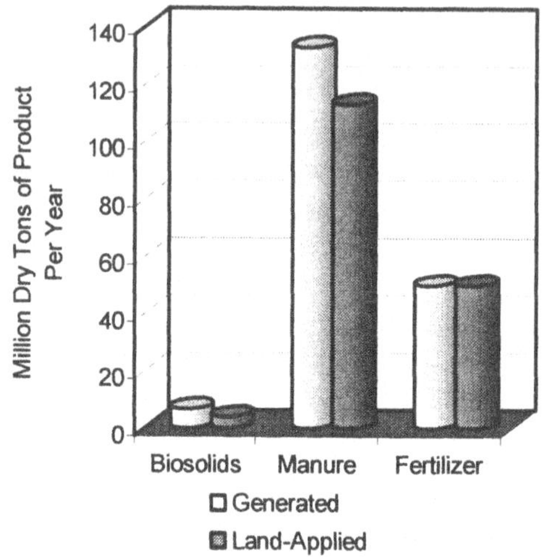

This chapter provides an overview of the products investigated for this project: what they are; how they are made; why they are used; how they are regulated; and what concerns have been raised regarding their use. This chapter also explains the approach that was used to collect and prepare the information presented in this report.

1.2 Study Approach

To prepare this document, the project team used a multistep approach that included the following activities:

1. ***Confirm Target Areas of Investigation.*** Because a goal of this project is to address publicly raised questions, the project team conducted a small, random survey of agricultural professionals to ensure that project research reflected key questions regarding amendment use.
2. ***Collect Relevant Literature.*** A comprehensive literature review was the foundation of this project. Sources reviewed for this document included:
   - Journals and texts identified using major databases such as Agricola, Biosis, and Medline;
   - Documents published by U.S. Environmental Protection Agency (USEPA) and the U.S. Department of Agriculture (USDA);
   - Gray literature (conference proceedings, articles in non-peer reviewed technical journals, technical reports and similar materials); and
   - Previous WERF studies.
3. ***Assess Relevant Literature.***
4. ***Identify Research Needs.*** The review of existing literature identified gaps in comparative data. For example, while there is a great deal of information on metals content in

biosolids, little work in this area has been done on manures. Chapter 5 presents research needs identified during this project.

The remaining portion of this chapter summarizes the findings of the project team.

1.3 Organic Soil Amendments

Organic soil amendments used in agriculture include biosolids, to a lesser extent, municipal solid waste (MSW) and yard waste composts, animal manures and crop residues. Although these products have some nutrient value and are sometimes used as fertilizers, their organic content is what differentiates them from mineral fertilizers (for the purpose of this document). With the exception of crop residues (and, until recently, animal manures) amendments have historically been viewed as waste products, although their critical role in recycling nutrients in the environment and replenishing soil organic matter has long been recognized.

With respect to organic products, this document focuses on amendments commonly used in agriculture that are currently viewed as waste products in the United States (manures, biosolids, and MSW and yard waste composts). Crop residues (or green manures) are, therefore, not addressed in this report.

Other organic products, such as paper mill sludges, cotton burr composts, and other materials are also used for agricultural purposes, but because of their relatively low volume use, they are not addressed here.

1.3.1 Biosolids

Biosolids are the recyclable products of sewage sludge that have been processed to reduce their pathogen content and vector attraction. Although biosolids are derived from sewage sludge, they are not directly comparable to sludge because of the intensive processing that they have undergone; simply stated, biosolids are to sewage sludge what gasoline is to crude oil.

Comprehensive federal regulations mandate the level of processing required for biosolids to be used in land application programs, as well as their pollutant contents (Box 1-1). These regulations — 40 Code of Federal Regulations (CFR) 503, Standards for the Use or Disposal of Sewage Sludge — were promulgated in 1993 and have undergone minor changes since that time. The rule may undergo further changes in the near future as a result of a current proposal (USEPA, 1999b) to add dioxin to the list of regulated compounds.

A variety of processes are used to convert raw sewage sludge into biosolids, and they are generally classified according to their ability to reduce pathogens (Box 1-1). Some common processes used to create biosolids include:

Box 1-1: 40 CFR 503 Standards for the Use or Disposal of Sewage Sludge

The land application of sewage sludge is regulated on the federal level by the 40 CFR 503 *Standards for the Use or Disposal of Sewage Sludge*. Promulgated in 1993, the 40 CFR 503 ("the 503 rule"), established nationwide criteria governing the land application, surface disposal and incineration (in sludge-only incinerators) of municipal sewage sludges. Standards established include pollutant limits, management practices and operational criteria, as well as monitoring, record keeping and reporting requirements for sludge use and disposal. For land application, the rule establishes metal limits, pathogen reduction requirements, and vector attraction reduction requirements. As shown in Table 1-1, the 503 rule provides four sets of metal criteria for biosolids use. The rule also establishes two classes of pathogen reduction — Class A and Class B (Table 1-2).

Table 1-1
40 CFR Part 503 Metal Criteria for Biosolids Land Application

Pollutant	Ceiling Concentration *mg/dry kg*	Cumulative Pollutant Loading Rates (CPLR) *kg/ha*	Pollutant Concentration *mg/dry kg*	Annual Pollutant Loading Rates (APLR) *kg/ha*
Arsenic	75	41	41	2
Cadmium	85	39	39	2
Chromium	-	-	-	-
Copper	4300	1500	1500	75
Lead	840	300	300	15
Mercury	57	17	17	1
Molybdenum	75	-	-	-
Nickel	420	420	420	21
Selenium	100	100	100	5
Zinc	7500	2800	2800	140

Table 1-2
Criteria for Meeting Class A and B Requirements

Parameter	Unit	Limit
Criteria for Meeting Class A Requirements		
Fecal Coliform	MPN/g TS(1)	1,000
Salmonella	MPN/4g TS	3
And ***one of the following process options:***		
Temp/Time based on % Solids	Alkaline Treatment	
Prior test for Enteric Virus/Viable Helminth	Post test for Enteric Virus/Viable Helminth	
Composting	Heat Drying	
Heat Treatment	Thermophilic Aerobic Digestion	
Beta Ray Irradiation	Gamma Ray Irradiation	
Pasteurization	PFRP(2) Equivalent Process	
Criteria for Meeting Class B Requirements		
Fecal Coliform	MPN or CFU/g TS(3)	2,000,000
Or ***one of the following process options:***		
Aerobic Digestion	Air Drying	
Anaerobic Digestion	Composting	
Lime Stabilization	PSRP(4) Equivalent	

(1) Most probable number per gram dry weight of total solids.
(2) Process to Further Reduce Pathogens.
(3) Most probable number or colony-forming units per gram dry weight of total solids.
(4) Process to Significantly Reduce Pathogens.

Because of the higher pathogen content of Class B biosolids, the 503 rule also imposes the use restrictions (such as harvesting, access, and grazing restrictions) on these products. Vector attraction reduction criteria must be met for all biosolids products, but the rule presents 12 options to meet this goal, as shown in Table 1-3.

Table 1-3
Summary of Requirements for Vector Attraction Reduction Options

Option	Requirement	Where/When Requirements Must Be Met
1 Volatile Solids (VS) Reduction	> 38% VS reduction during solids treatment	Across the process
2 Anaerobic bench-scale test	< 17% VS loss, 40 days at 30°C to 37°C (86°F to 99°F)	On anaerobic digested biosolids
3 Aerobic bench-scale test	< 15% VS reduction, 30 days at 20°C (68°F)	On aerobic digested biosolids
4 Specific Oxygen Uptake Rate (SOUR)	SOUR at 20°C (68°F) is < 1.5 mg oxygen/hr/g total solids	On aerobic stabilized biosolids
5 Aerobic Process	> 14 days at > 40°C (104°F) with an average> 45°C (113°F)	On composted biosolids
6 pH adjustment	> 12 measured at 25°C (77°F)*, and remain at pH > 12 for 2 hours and > 11.5 for 22 more hours	When produced or bagged
7 Drying without primary solids	> 75% Total Solids (TS) prior to mixing	When produced or bagged
8 Drying with primary solids	> 90% TS prior to mixing	When produced or bagged
9 Soil Injection	No significant amount of solids is present on the land surface 1 hour after injection. Class A biosolids must be injected within 8 hours after the pathogen reduction process.	When applied
10 Soil Incorporation	< 6 hours after land application; Class A biosolids must be applied on the land within 8 hours after being discharged from the treatment process	After application
11 Daily cover at field site	Biosolids placed on a surface disposal site must be covered with soil or other material at the end of each operating day	After placement
12 pH adjustment of septage	> 12 measured at 25°C (77°F)*, and remain at pH > 12 for 30 minutes without addition of more alkaline material.	Septage

* or corrected to 25°C
Source: 40 CFR 503

To encourage the beneficial use of biosolids , the 503 rule contains two regulatory schemes, depending on the quality of the biosolids in question. Biosolids that are proven to be of "exceptional quality" (i.e., Class A products that meet the most stringent metal criteria) are exempt from further regulatory controls and may be used as freely as other soil amendments and fertilizers. Other biosolids are subject to use limitations.

- ***Digestion:*** Digestion is used to reduce pathogens and vector attraction in liquid sludges. Generally, the liquid sludges are heated to temperatures of at least 95 degrees Fahrenheit (°F) or 35 degrees Celsius (°C), reducing their pathogens and volatile solids content. The resulting liquid biosolids are usually applied to the land as is or are dewatered before land application. Digested biosolids can either be Class A or Class B products (see Box 1-1), depending upon the temperature at which they were heated and the length of time during which they were processed.
- ***Composting:*** Composting is essentially an acceleration of natural degradation processes. Microbial activity heats the composting material, generating sufficient heat to destroy pathogens and dry the composting mix. The resulting product is humus-like, and can be used in agriculture, horticulture, or land reclamation. Composting is generally considered to be a Class A process.
- ***Heat-Drying:*** As its name implies, this process heats sludge to high temperatures (about 500°F to 900°F, 260°C to 480°C) killing pathogens and removing moisture. The resulting product is almost free of water (about 90% solids), and is either granular or formed into pellets about 1/8 inches (3 mm) in diameter. Heat-drying is considered to be a Class A process.
- ***Alkaline Stabilization:*** This process mixes alkaline materials such as lime with the sludge. A chemical reaction rapidly heats the mix, increases the pH, and dries the material. The resulting product has high pH (about 12.5 at the end of the process) and can be used as a liming agent. Variations of this process can be used to meet Class A or Class B pathogen reduction requirements.

1.3.2 MSW and Yard Waste Compost

Concerns about the rapidly diminishing landfill capacity in our country have led to the recycling of organic municipal solid waste (MSW). As Figure 1-2 shows, greater than 50% of the municipal wastestream, including food, yard waste, and paper, is organic and potentially suitable for composting.

In 1999, almost 400 million tons (363 million metric tons) of MSW were generated in the United States (Goldstein, 2000), but, a 1999 BioCycle survey revealed estimates of less than 500,000 tons (454,000 metric tons) of MSW were composted annually (Glenn and Block, 1999). Predominant markets for MSW compost at this time are nurseries, athletic fields, and residential uses, although agriculture has been cited as an existing and potential market for MSW compost.

As many cities and towns seek to reduce the amount of waste deposited in landfills, yard waste composting is becoming an accepted and growing practice. In 1990, an estimated 34 million tons (or 31 million metric tons) of yard trimmings were collected, about 12% of which were recycled (USDA, 1998). Today, we expect that both the quantity generated and the overall recycling rate have increased, but it is difficult to ascertain the exact quantity of recycling activity that is attributable to composting.

Figure 1-2
1995 Municipal Wastestream Components
(percentage by weight)

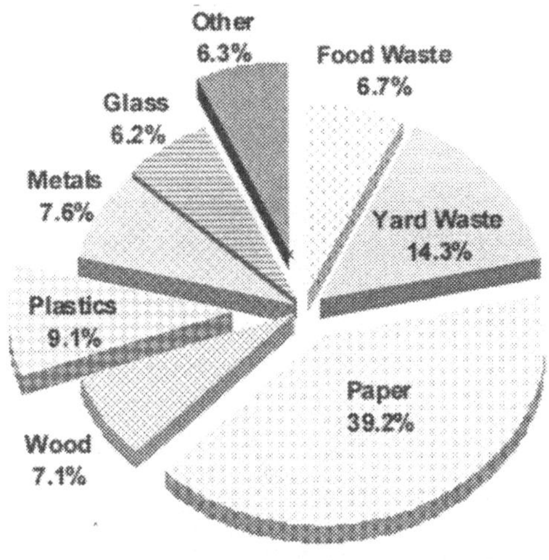

Source: USEPA, 1997a

Unlike biosolids, MSW and yard waste compost use is not regulated at the federal level. However, some states maintain compost quality guidelines or regulated standards for nonbiosolids composts.

1.3.3 Manures

Major manure sources in the United States include cattle, poultry, and swine (Figure 1-3). Most of these manures are land-applied in liquid, semisolid, or solid forms, supplying organics and nutrients to the soil. The characteristics of these materials vary widely, depending upon animal type, bedding practices, feed, animal age, environment, storage and treatment, and a host of other factors.

Figure 1-3
1992 Manure Generation Estimates[a] (Million tons)

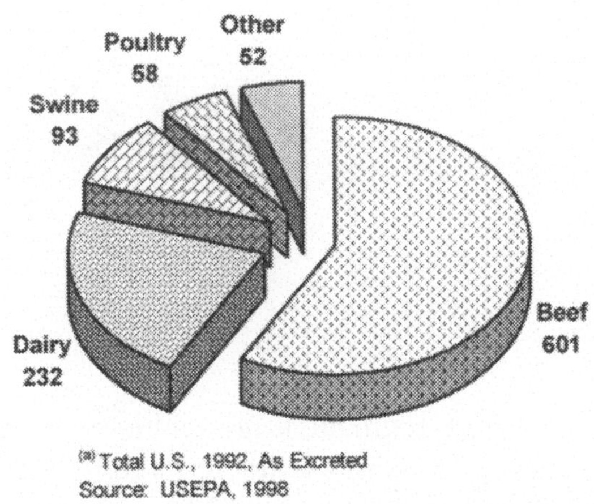

[a] Total U.S., 1992, As Excreted
Source: USEPA, 1998

As Figure 1-4 shows, these manure sources are unevenly distributed around the country, and clustered in certain areas. This clustering of livestock production mirrors an even stronger trend in this nation: the replacement of small family farms with large operations that house thousands of animals. Because manures are usually land-applied on farms or on nearby acreage, this concentration of livestock (both regionally and on larger farms) can yield an oversupply of nutrients and other environmental problems in some areas.

Figure 1-4

1997 Excess Manure Nitrogen and Phosphorus Assuming no Export of Manure from Farm

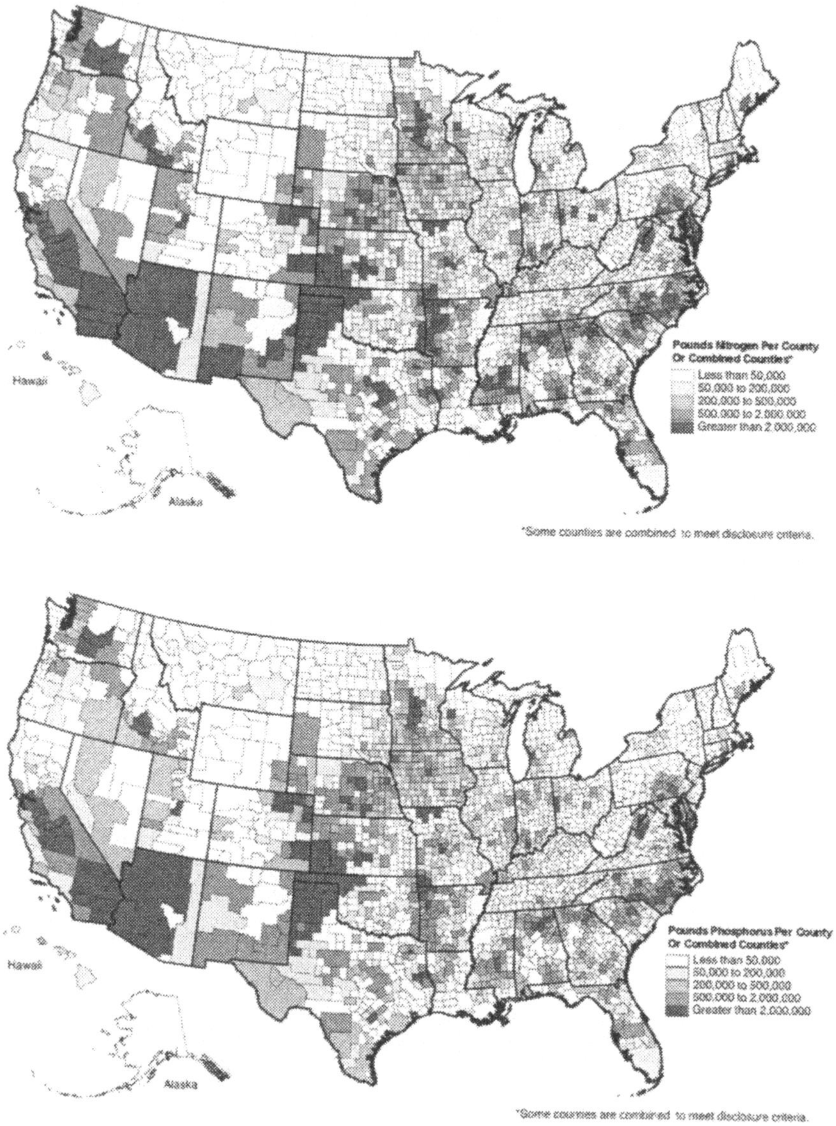

Source: Kellog, et, al., 2000

Until 1999, there were no federal regulations governing most manure applications, although many states have rules that address this issue for larger livestock facilities, or for animal feeding operations (AFOs). In 1999, the USDA and USEPA announced a Unified National AFO

Strategy to minimize water quality and public health impacts from AFOs and manure applications (USDA/USEPA, 1999).

Reflecting the goals stated in the Unified Strategy, the USEPA has recently proposed more stringent regulations (USEPA, 2001) for large AFOs, or Concentrated Animal Feeding Operations (CAFOs). CAFOs are currently governed by National Pollution Discharge Elimination System (NPDES) permits. Proposed changes in NPDES regulations include options for redefining CAFOs (based upon size and potential environmental impact) and additional permit requirements, but they also address manure application concerns. Critically, the CAFO definition has been expanded to include the manure land application area, as well as the production area, and manure applications are regulated accordingly. Each CAFO would be required to prepare a site-specific nutrient management plan to ensure that excess nutrients are not applied to the land. AFOs are not required to meet the NPDES requirements but, under the Unified Strategy, they are strongly encouraged to participate in voluntary comprehensive nutrient management planning.

To our knowledge, there are currently no maximum limits on specific constituents in manures. However, the American Association of Plant Food Control Officials (AAPFCO) has recently proposed that the metal limits contained in 40 CFR 503 also apply to manures and manipulated manures (that are not supplemented with chemical fertilizers) (AAPFCO, 2001). Additionally, at least one state has established metal limits in manure-amended soils. Concerns regarding zinc and copper phytoxicity from land-applied manures in North Carolina has led to limits for these parameters for fields where manures are applied (SB1217 Interagency Group, 2001). Manure cannot be applied to lands where zinc and copper concentrations have a Soil Test Index of 3000 (equivalent to 120 parts per million (ppm) and 20 ppm, respectively). A zinc soil test index of 500 (with a soil pH of 6) has been established for peanut crops.

1.4 Mineral Fertilizers

Fertilizers fall into three general categories based upon what they provide: primary nutrient fertilizers; secondary nutrient fertilizers; and micronutrient fertilizers. Primary nutrient fertilizers provide one or more of the key plant nutrients nitrogen, phosphorus, and potassium.

Nitrogen fertilizers, including ammonium nitrate, urea, nitrogen solutions, ammonium sulfate, and ammonium phosphate, supply inorganic nitrogen for plant growth. Most are derived from anhydrous ammonia, which can also be directly applied to the land.

The nutrient content of phosphorus fertilizers is actually measured in terms of available phosphorus pentoxide (P_2O_5), or phosphate. Phosphate rocks are the primary source for phosphate fertilizers, which include superphosphates and diammonium phosphates. The rocks are treated with sulfuric acid or phosphoric acid to produce monocalcium phosphate, a primary phosphorus fertilizer component.

The nutrient value of potassium fertilizers is defined by potassium oxide (K_2O), or potash content. Potash fertilizers are derived from potash ores that can be used directly as fertilizer. These ores are often processed, however, to remove impurities such as salt (TFI, 1982).

As Figure 1-5 shows, between 1960 and 1980, nitrogen fertilizer use increased dramatically before leveling-off somewhat in recent years. Phosphate and potash use has remained relatively steady over the last several decades. The rapid increase in nitrogen use was attributed to market demand created by favorable crop responses, mainly with corn (USDA, 1997).

Figure 1-5
U.S. Commercial Fertilizer Use, 1960-95

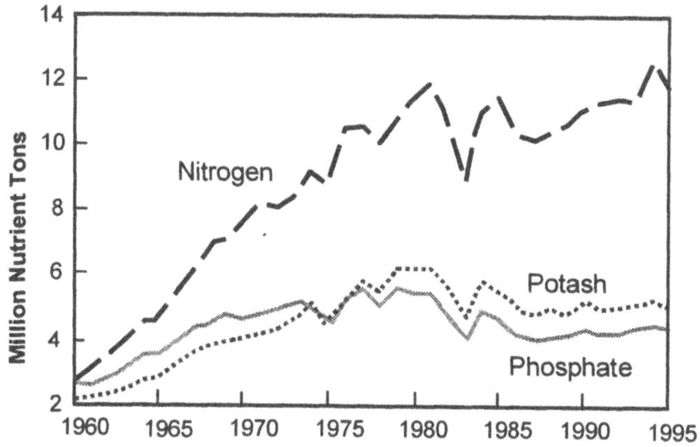

Source: USDA, 1997

Secondary nutrient fertilizers provide one or more of the nutrients calcium, sulfur and magnesium, but these nutrients are required in lesser amounts than primary nutrients. Calcium sources include gypsum (calcium sulfate), calcitic and dolomitic limestones, and other liming materials. Dolomitic limestone is the most common source of magnesium (USDA, 1997). Sulfur sources include gypsum, sulfuric acid, and sulfur supplied by organic sources (USDA, 1997).

Micronutrient fertilizers provide one or more of the essential elements boron, chlorine, copper, iron, manganese, molybdenum and zinc, although plants require just trace amounts of micronutrients. Materials used to manufacture micronutrient fertilizers vary, but can include industrial by-products such as steel flue dust or cement kiln dust. Some of the raw materials in the manufacturing process, until they are recycled as fertilizer, are classified as hazardous waste. Historically, the regulation of fertilizers focused on ensuring that the nutrient content of the product met label claims. Many states have such requirements, and though no federal standards exist, the Association of American Plant Food Control Officials (AAPFCO) has aimed at ensuring uniformity among state regulations.

Before 1997, contaminant levels in fertilizers, except for waste-derived products, were not regulated at the federal or state level in the United States (WSDE, 1999). Fertilizer metals have been regulated in Canada for many years, however, and the Canadian standards are presented in Table 1-4. Note that these standards also apply to biosolids land-applied in Canada. However, concerns regarding the metals content of waste-derived fertilizers, and some phosphate fertilizers, led to the state of Washington's adaptation of the Canadian standards for use in Washington. In Washington's 1998 rule (WAC-16-200-7064), metal limits are presented as a maximum annual application rate. They were converted from the cumulative loads in the Canadian standards to an annual rate by assuming an application period of 45 years (Alexander and Goldstein, 1999).

Table 1-4
Canadian Standards

Metal	Maximum Acceptable Cumulative Metal Addition to Soil (kg/ha)
Arsenic	15
Cadmium	4
Cobalt	30
Copper	100
Mercury	1
Molybdenum	4
Nickel	36
Selenium	2.8
Zinc	370

Source: Alexander and Goldstein, 1999

Texas, California, and Utah also control metal concentrations in fertilizers, albeit all on a different basis. For example, the Texas Administrative Code (Title 4, Part 3, Chapter 65) requires a warning label to be affixed to any fertilizer that, if used in accordance with label instructions, exceeds an allowable metal loading rate. The rate is based upon limits established under 40 CFR 503 (see Box 1-1). The California Department of Food and Agriculture (CDFA) has proposed risk-assessment-based limits for arsenic, cadmium, and lead. (Risk assessments for fertilizers are discussed in Section 4 of this document.) Utah requires that potential distributors provide a metal analysis of products that they intend to distribute within the state. To date, they have used the Washington Standards for Metals as a basis of comparison, but they may adopt national standards currently proposed by AAPFCO when those standards are finalized.

AAPFCO has proposed metal limits (Table 1-5) based upon risk assessments performed for The Fertilizer Institute (The Weinberg Group, 2000) the USEPA (USEPA, 1999c), and the state of California (Foster Wheeler, 1998).

Based upon concerns regarding waste-derived zinc fertilizers, USEPA is also proposing to set limits on the concentrations of some metals in these materials (USEPA, 2000b). Proposed limits are shown in Table 1-6. In its proposal, the USEPA is also requesting comments regarding whether it should also address concerns about metals in iron fertilizers from mined materials.

Table 1-5
Proposed AAPFCO
Metal Limits for Fertilizers[1]

Metal	*ppm per 1% Phosphate*	*ppm per 1% Micronutrients[2]*
Arsenic	13	112
Cadmium	10	83
Cobalt	3,100	23,000
Lead	61	463
Mercury	1	6
Molybdenum	42	300
Nickel	250	1,900
Selenium	26	180
Zinc	420	2,900

Source: American Association of Plant Food Control Officials (AAPFCO), 2001

Notes:

(1) Applies to fertilizers that contain guaranteed amounts of phosphates and/or micronutrients, with the exception of composts.

(2) Micronutrients include iron, manganese, zinc, copper, molybdenum, and boron. Cobalt and selenium can also be considered to be micronutruents.

Table 1-6
EPA Proposed Metal Limits for Zinc Fertilizers Made from Recycled Hazardous Secondary Materials

Metal	*Maximum Allowable Total Concentration* *ppm per unit (1%) of zinc*
Lead	2.8
Cadmium	1.4
Arsenic	0.6
Mercury	0.3
Nickel	1.4
Chromium	0.6

Source: USEPA, 2000b

1.5 Overview of Benefits and Concerns

About 20% of the land in the contiguous United States is devoted to the production of crops (USDA, 1997). Without the addition of nutrients to the soil, the United States could not sustain its current high crop yields. Both fertilizers and, to a lesser extent, soil amendments, supply the nutrients necessary to support crop growth. Soil amendments can also replenish the valuable organic matter that erosion and some farming practices have depleted.

Our role as a leading crop producer, however, has not come without an environmental price. As traditional farming practices and erosion have removed topsoil, the organic matter of our soils has fallen steadily. Before cultivation, the average organic content of the nation's soils was about 3 to 5% (Logan, 2001). A recent study of 193 soils showed that the average organic matter in US agricultural soils has fallen to about 1.4% (Holmgren et al., 1993). With the adoption of conservation tillage practices, where crop residues are returned to the soil and less tillage is used, indications are that this steady decline is reversing (Allmaras et al., 2000), but the changes are small (Logan, 2001).

Agriculture has also been cited as a leading cause of impairment of our nation's rivers and lakes (USDA, 1997). To date, the water quality of about 23% of our nation's rivers and streams, 42% of its lakes and 32% of its estuaries has been assessed (USEPA, 2000c). Although a majority of the assessed water bodies are in good health, others are impaired. In impaired rivers and streams, agriculture is considered to be a leading source of pollution (agriculture is not, however, a leading pollutant contributor to the nation's estuaries, based upon water quality data collected to date) (USEPA, 2000c). Agriculture has also impacted groundwater in some areas. A survey of drinking water wells revealed that rural wells were twice as likely to have high nitrate levels as urban wells (Sharpley et al., 1998).

Concerns have been raised regarding the nutrient impacts from both fertilizers and soil amendments, and the safety of biosolids and manures. Depending on the type of product, these impacts can include the accumulation of metals in soils, plant uptake of land-applied metals, the contamination of groundwater and soil with organic compounds, odors, and potential health impacts from pathogens and viruses.

This document subsequently explores these issues in greater depth.

1.6 Nutrients and the Environment

Many of the benefits and concerns regarding the products investigated here focus on the nutrients nitrogen and phosphorus. Accordingly, it is helpful to understand nature's cycling of these nutrients.

1.6.1 The Nitrogen Cycle

Nitrogen is ubiquitous in nature, and elemental nitrogen (N_2) makes up about 78% of the air that we breathe. Nitrogen used by crops, however, must be in or converted to a *bioavailable* form: that is, a form which plants can absorb and process. Plant-available forms of nitrogen include ammonium (NH_4) and nitrate (NO_3), both of which are inorganic.

Figure 1-6 shows the nitrogen cycle and the many ways that nitrogen is transformed and transits through the environment.

Figure 1-6
The Nitrogen Cycle

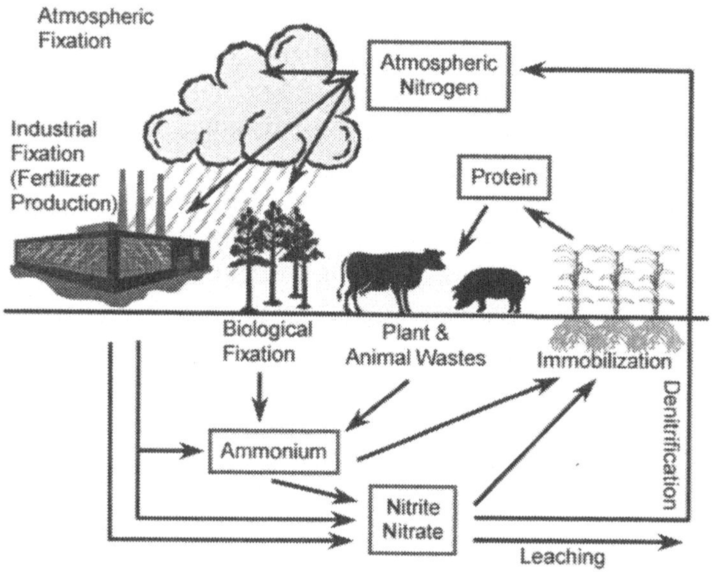

Adapted from: O'Leary et al., 1997

Nitrogen can be *fixed* (converted from nitrogen gas to an inorganic and bioavailable nitrogen form) through biological processes (e.g., symbiotic fixation by legumes) or industrial processes (e.g., fertilizer production). It can also be fixed by atmospheric phenomena, such as lightening, adding possibly 10 to 20lb/ac (1.12 to 22.4 kilogram per hectare (kg/ha)) to soil annually (TAES, 2001).

Two nitrogen-transformation processes that should be in balance are *mineralization* and *immobilization*. Mineralization is the microbial transformation of organic nitrogen to inorganic nitrogen through the decomposition of organic matter. Since most nitrogen in the soil is organic, this is a critical process. Immobilization refers to the adsorption of inorganic nitrogen by plants and microorganisms, and its transformation into amino acids and proteins.

If the amount of nitrogen made available to plants through mineralization or the addition of nutrients (amendments, fertilizers or crop residues) exceeds the needs of the plants, then excess nitrogen, particularly as nitrate, can leach to groundwater.

Not all nitrogen in the soil is plant-absorbed or leached from the soil, however. *Denitrification* (the biological transformation of nitrate to nitrogen gas) and *volatilization* (the loss of ammonia from the soil as a gas) can reduce the amount of plant-available nitrogen in a soil.

1.6.2 The Phosphorus Cycle

Figure 1-7 illustrates the phosphorus cycle as presented in the USDA's document "Agricultural Phosphorous and Eutrophication" (Sharpley et al., 1999).

Figure 1-7
The Phosphorus Cycle in Soil

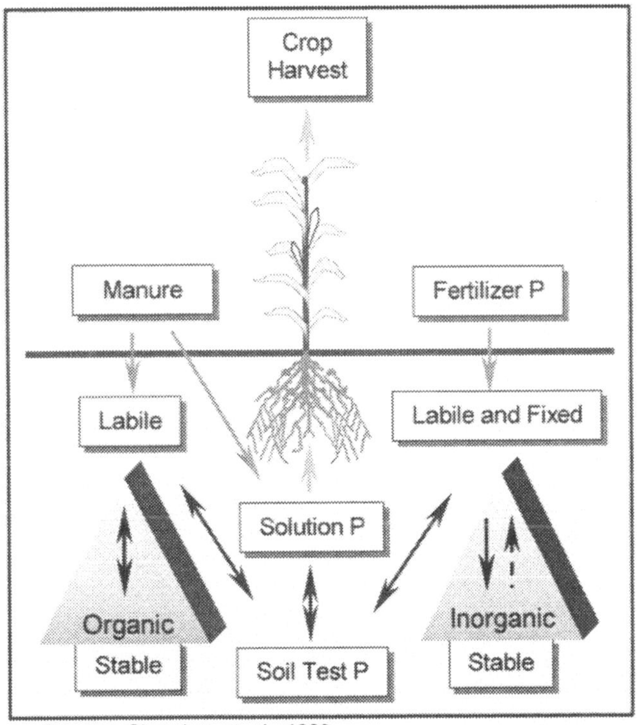

Source: Sharpley, et. al., 1999

Like nitrogen, soil phosphorus is present in organic and inorganic forms, but phosphorus (and its interactions in the soil) is different from nitrogen in many respects. It is not present in its elemental state in nature, and differences between organic and inorganic forms are not clear-cut (Sharpley et al., 1999). Soil organic phosphorus includes undecomposed residues, microbes, and organic matter. The organic phosphorus is gradually transformed from stable to labile forms that are available to plants. These inorganic phosphorus forms are used by plants, but vary in both solubility and availability. Phosphorus can also bind with calcium, iron or aluminum (especially in acid soils) and become less available. Because it binds with so many different compounds (and with clay in soils), phosphorus is relatively immobile, thus requiring management practices that are unlike those used for nitrogen. Chapter 4 discusses the impact of phosphorus from agricultural inputs on the environment.

CHAPTER 2.0

PRODUCT CHARACTERISTICS

Soil amendments and fertilizers are similar in some respects and different in other respects. To compare the impacts of these products on the environment, we must first understand how the products themselves compare. This chapter examines amendment and fertilizer characteristics and compares them, where data is available to do so.

This chapter is divided into six sections: (1) physical characteristics; (2) nutrient content and availability; (3) metal content and availability; (4) biological characteristics; (5) other chemicals; and (6) odor potential. Each subchapter discusses the properties of biosolids products, MSW and yard waste composts, manures, and mineral fertilizers.

2.1 Physical Characteristics

Physical characteristics of concern for agricultural applications include texture, particle size (for ease of spreading) organic content, solids content, pH and bulk density. *Volatile solids content,* a measure of the organic, biodegradable fraction of an amendment's total solids content, may also be important, as it can be a measure of degradation (and odor) potential.

2.1.1 Biosolids Products

Biosolids' physical textures and particle sizes are as varied as the processes that are used to create them. Heat-dried biosolids can be granular, but more often, are spherical and similar in size to a small ball bearing. Of the biosolids discussed here, heat-dried products are closest in appearance to mineral fertilizers; in fact, they can be and have been incorporated into fertilizer blends. Composts are similar in texture to peat, but depending upon how they are made, they may contain some wood fibers or wood chips. Alkaline stabilized materials can have a clay-like or a soil-like texture, depending upon their moisture content and other factors. Digested biosolids can be applied to land in liquid form, or as a clay-like dewatered material. Table 2-1 presents some bulk densities, organic contents, total solids contents, and volatile solids contents for these products.

2.1.2 MSW and Yard Waste Composts

To some extent, in appearance and texture, these composts are similar to biosolid composts. However, both MSW and yard waste composts may contain *inerts*, pieces of plastic, glass, etc., depending upon how the compost *feedstocks* (materials composted) were handled before the composting process. For example, shredding the waste before composting, which is often required to reduce large materials to smaller, compostable pieces, can tear plastics into small sheets that remain in the product (New and Papworth, 1988).

Table 2-1
Typical Biosolids Physical Properties

Property	*Alkaline Stabilized (Class A)*	*Compost*	*Alkaline & Digested Liquid*	*Alkaline & Digested Cake*	*Heat-dried*
Total Solids (%) (1)	55	58	4	22	95
Volatile Solids (% of total solids) (1)	60	60	61	60	67
Bulk Density (lb/cf)	65-75	45	60	55	33-46 (2)
Organic Content (%)	12(3)	50-75(3)	60-75	60-75	78(3)
pH					6.4-8.0(2)

(1) Lue-Hing et al., 1998

(2) Grace, Carr and Fox (1994)

(3) CDM, 1998

2.1.3 Manures

The forms of land-applied manure vary as much as the forms of biosolids used in agriculture. Manures are usually applied to the land as a liquid (0% to 10% solids range); a semisolid (10% to 20% solids content); or a solid (20% to 25% solids content). However, solids contents can be 50% or greater where manures are stockpiled before they are applied to the land or — as is the case for poultry litter — before wood shavings or other dry materials are added. Additionally, manure composts can be used for agricultural purposes. Relatively little data on manure composts is available at this time. This could be because manure composting, to date, has received far less attention than biosolids, MSW, or yard waste composting.

Physical properties of manures can vary widely depending upon animal type and age; whether or not bedding is included in the manure; handling procedures; and other parameters (Midwest Plan Service [MWPS], 1993). Many land grant universities and other institutions have compiled data on manure characteristics (e.g., MWPS, 1993; Natural Resources Conservation Service [NRCS], 1999). Most note that actual physical properties may vary significantly from the typical values that they present. Table 2-2 presents data collected from some of these sources.

Table 2-2
Animal Waste Physical Characteristics, as Excreted

Animal	Total Solids Concentration (%)	Volatile Solids Fraction of TS (% of TS)	pH	Bulk Density (lb/cubic foot)
Dairy cattle	14	83	7.0	62
Beef cattle	15	85	7.0	63
Swine	13	77	7.5	62
Poultry				
Layers	25	75	6.9	60
Broilers	26	77	na	63

Source: American Society of Agricultural Engineers Standards (ASAE), 2000. Manure Production and Characteristics. ASAE D384.1 DEC 99. Total solids and volatile fraction computed from that data. Reprinted from the ASAE Standards with permission of ASAE

na = Not yet identified

Manure organic matter is estimated to be 80% to 90% of dry weight (composted poultry litter is generally lower, about 60%). Organic matter content depends upon feed, animal metabolism, and storage and handling procedures (CAST, 1996). Little information is available for manure composts, but Barlow (unpublished) reports an organic content of about 44 to 63% for composted turkey droppings and pine bark bedding.

2.1.4 Mineral Fertilizers

Mineral fertilizers are applied to the land as solids, liquids, or suspensions (crystals of plant nutrients suspended in solutions). Solid fertilizers are usually in granular or pellet form, making them easily spreadable by conventional fertilizer spreaders. The physical characteristics of each type of fertilizer product are relatively uniform and will not be detailed here.

An important difference between soil amendments and mineral fertilizers is that fertilizers (with the exception of urea) contain no organics. Some fertilizer blenders have replaced filler material in fertilizers with pelletized biosolids, however, adding organics to these blends. Additionally, researchers and manufacturers are exploring the blending of mineral fertilizers and sewage sludge before drying (Burnham, 1999; Ireland and Thayer, 2000). These innovative hybrid products have yet to be used extensively in agriculture, thus, they are not discussed here.

2.2 Nutrient Content and Availability

To a large extent, literature identified through this study focused on the total *NPK,* or primary nutrient content, of soil amendments and fertilizers. NPK is an acronym that reflects the elemental symbols for nitrogen (N), phosphorus (P) and potassium (K), but actually expresses the nitrogen (N), phosphate (P_2O_5), and potash (K_2O) content of a fertilizer or soil amendment as a percentage.

Agriculturists consider the availability of nutrients in amendments of critical importance, but this subject has yet to be satisfactorily researched for some products (especially manures). Most research to date (on manures and other organic amendments) has focused on nitrogen, rather than on phosphorus.

In addition to primary nutrients, the following text addresses secondary nutrients and some micronutrients. Chapter 2.3 also covers micronutrients, especially those considered to be heavy metals.

2.2.1 Biosolids

Because many biosolids are used as soil amendments and/or fertilizers, their primary nutrient content is often reported. Table 2-3 presents primary nutrient data collected for various biosolids products. As the table shows, most biosolids have a low potassium content. One exception to this may be some Class A alkaline stabilized products: the high mica content of some alkaline additives can raise the potassium content to 1% to 3% (Yamakawa, 2000).

Secondary nutrient data are less available. Table 2-4, however, shows the secondary nutrient concentrations for some biosolids.

Table 2-3
Biosolids Primary Nutrient Contents

Nutrient Content (as % of dry weight)	*Alkaline Stabilized (Class A)*	*Compost*	*Alkaline & Digested Liquid*	*Alkaline & Digested Cake*	*Heat-dried*
Nitrogen	1.0	2.8	5.3	4.1	6.0
Total Phosphorus	0.4	1.7	2.2	1.9	3.1
Potassium	0.3	0.3	0.3	0.2	0.3

From: Lue-Hlng et al., 1998

In general, Table 2-3 illustrates the considerable variability in the nutrient content of different biosolids products. Even within a single type, nutrient content can vary depending on the wastewater source, treatment process, and even sludge processing approaches. Table 2-5, for example, shows the impact of digestion and dewatering process on biosolids nutrient content. The volatilization of ammonia is an important mechanism for nitrogen loss.

Table 2-5 also illustrates a key concept with regard to biosolids and nitrogen: most of the nitrogen in biosolids is in an organic form, which must be mineralized before it is available to plants. Figure 2-1 provides a breakdown of the nitrogen content for select biosolids products.

Table 2-4
Biosolids Secondary Nutrients

Nutrient	Range (%)	Source
Calcium		
	1.0-4.0	Sullivan, 1998 (1)
	2.7-4.8	Linden et al., 1995 (2)
Magnesium		
	0.4-0.8	Sullivan, 1998 (1)
	0.6-0.8	Linden et al., 1995 (2)
Sulfur		
	0.6-1.3	Sullivan, 1998 (1)
	NA	

(1) Data from 7 Pacific Northwest plants; composts and lime-stabilized materials not included.
(2) Reflects 20 years of data for five liquid sludges in Minnesota

Table 2-5
Impact of Dewatering and Digestion on Nutrients

Nutrient	Raw Sludge	Digested Biosolid No. 1 (1)	Digested Biosolid No. 2 (2)
Primary Nutrients			
Total Nitrogen, %	5.5	4.7	3.7
Ammonia, %	0	0	0
Organic Nitrogen, %	5.4	4.7	3.7
Nitrate/nitrite, mg/dry kg	0	0	0
Total Phosphorus, %	2	2.3	2.6
Available Phosphate, %	0.1	0	0
Potassium, %	0.3	0.2	0.1
Secondary Nutrients			
Calcium, %	4.2	6.9	14
Magnesium, %	0.3	0.3	0.4
Sulfur, %	1.1	0.9	1
Micronutrients			
Boron, mg/dry kg	<5	7.6	159
Iron, %	4.1	4.2	4.4

Adapted from: Grace, Carr and Fox, 1994
(1) belt press dewatered
(2) mix of belt press and plate and frame dewatered

Nitrogen mineralization is a gradual process, and that is why the nitrogen in biosolids is sometimes referred to as "slow release" — the inorganic nitrogen in the biosolids is immediately available to plants, whereas the organic nitrogen is slowly released to them.

To date, it has been difficult to predict the rate at which organic nitrogen becomes available for plants (i.e., the mineralization rate). Many researchers have studied mineralization rates, but most researchers have used methods that were not practical or were too expensive for agricultural professionals to use daily. USEPA provides a table of nitrogen mineralization rates for different biosolid products (USEPA, 1993), but other studies have shown a considerable variance from USEPA numbers (Camp Dresser & McKee [CDM] Inc., 1998). Estimating the mineralization rate, however, is critical to achieving the dual objectives of supplying crop nutrient needs and protecting the environment (from nitrogen leaching).

Figure 2-1

Organic and Inorganic Nitrogen Fractions in Biosolids [1]

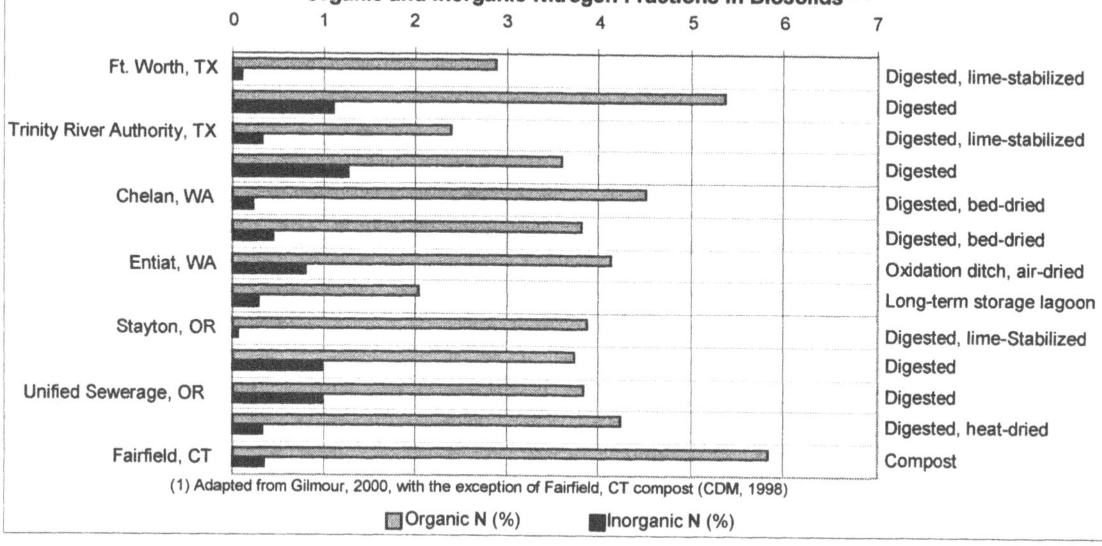

(1) Adapted from Gilmour, 2000, with the exception of Fairfield, CT compost (CDM, 1998)

A recent WERF study (Gilmour, 2000), however, focused on developing improved methods to estimate plant available nitrogen (PAN). This study reached these conclusions:

- PAN is directly related to biosolids total nitrogen content.
- First and second year PAN can be estimated using a constant mineralization rate for a given location.
- Large amounts of ammonia can volatilize from surface-applied biosolids.

The authors have developed a simple, four-step process to estimate PAN. The new approach, which relies on standard and inexpensive laboratory tests, should be appropriate for most biosolids. The approach may not be suitable, however, for biosolids with a *carbon-to nitrogen ratio* (total carbon content divided by total nitrogen content) of 15 or less. The study showed, however, that many different biosolids types should meet this C:N criterion. One exception might be composts, if they have not been screened well enough to remove woodchips or have not been sufficiently composted to achieve higher C:N ratios.

New research is also beginning to yield information on how different wastewater treatment and stabilization processes impact phosphorus and its availability. In a study assessing phosphorus (P) availability as defined by water extractable P (WEP), Brandt, Elliot and O'Conner (2001) found that WEP in 18 biosolids samples was far lower than manure or fertilizer WEP, and ranged from <0.5% to 13.5%. The researchers found average WEPs for alkaline-stabilized, composted, aerobically treated, anaerobically treated, and heat-dried biosolids were 5.3%, 2.9%, 2.7%, 2.2%, and 0.5%, respectively. Unstabilized biosolids from biological nutrient removal processes exhibited a WEP of 13.5%. The researchers attribute the relatively low WEP in biosolids to the removal of water soluble P during dewatering and/or the addition of iron or alum for phosphorus removal. Water treatment plant residuals, which are often discharged to local WWTPs, can contain high concentrations of iron or aluminum, and can immobilize P as well (Peters and Basta, 1996). Sims et al. (1999) also noted that the addition of iron salts (for odor control, improved solids settling, or reduction of effluent phosphorus) is likely to immobilize biosolids P.

2.2.2 MSW and Yard Waste Composts

Little information exists on nutrient content and nutrient availability for these products. Based upon limited studies, it appears that MSW compost contains approximately 0.5% - 1.6% nitrogen; 0.14% to 0.35% phosphorus; and 0.28% to 0.76% potassium (Ozores-Hampton, 1994; Maynard, 1993; Golueke, Diaz, and Gurkewitz, 1989). The nitrogen content of yard waste compost is estimated to be 0.5% (Maynard, 1993), although higher concentrations have been reported. Harrison and Henry (1992) reported the nutrient content of yard waste compost to be 1.5% nitrogen, 0.4% phosphorus, and 0.7% potassium, while Golueke et al (1989) reported it to be 0.63%–0.90% nitrogen, 0.14%–0.16% phosphorus, and 0.62%–0.72% potassium.

With respect to secondary nutrients, Golueke, Diaz, and Gurkewitz (1989) noted that sulfur concentrations in MSW composts were greater than corresponding concentrations in yard waste or biosolids composts. For example, the sulfur content of an MSW compost was reported to be 1,000 parts per million (0.1%), whereas the yard waste and sulfur concentration was reported to be 0.2% to 0.26%.

No specific studies of nitrogen mineralization rates for nutrients in MSW or yard waste compost have yet to be identified, probably because these two materials are not considered large sources of nutrients.

2.2.3 Manures

Like biosolids nutrient content, manure nutrient content has been well documented, but nutrient availability has not. Simplified methods are being developed for assessing nutrient availability in manures, though.

Table 2-6 summarizes manure nutrient content data. It is important to note, however, that nutrient contents of manures vary considerably. Actual nutrient contents can vary from 20% (Veenhuizen et al., 1992) to 30% (MWPS, 1993) from the values presented in Table 2-6. Manure nutrient content varies by type and size of animal (Veenhuizen et al., 1992), as well as by feed, housing, bedding, moisture content, and so forth. The nitrogen fertilizer value of land-applied manures can be much less than that of fresh manures, because of losses (through ammonia volatilization) during storage and handling. Table 2-7 presents the estimated nutrient content of manures from various solids and liquid handling systems. Site-specific sampling and analysis are recommended to determine manure nutrient content because of myriad influencing factors.

Table 2-6
Nutrient Concentrations in Animal Waste

Animal	Nitrogen	Phosphorus	Potassium	Calcium	Magnesium	Sulfur	Sodium	Chloride	Iron	Manganese (mg/kg)	Boron (mg/kg)
Dairy cattle	3.8	0.8	2.4	1.3	0.6	0.4	0.4	1.1	0.1	158	59
Beef cattle	4.0	1.1	2.5	1.6	0.6	0.5	0.4	na	0.1	141	104
Swine	4.7	1.6	2.6	3.0	0.6	0.7	0.6	2.4	0.1	173	282
Poultry											
Layers	5.3	1.9	1.9	8.1	0.9	0.9	0.6	3.5	0.4	381	113
Broilers	5.0	1.4	1.8	1.9	0.7	0.4	0.7	na	na	na	na

Note: All values in % of dry weight unless otherwise noted.
Adapted from:ASAE, 2000. Manure Production and Characteristics,D384.1 DEC 99

na = Not yet identified

Table 2-7
Nutrients in Stored and Handled Manures (%, Dry Weight Basis)

Animal	Bedding or Litter	Solids Content	Ammonium N	Total N	Phosphate	Potash
Solid Manure						
Swine	No	18	1.4	3.1	2.2	1.4
	Yes	18	1.1	2.5	1.9	1.9
Beef	No	15	1.3	3.7	2.3	3.3
	No	52	0.7	2.0	1.3	2.2
	Yes	50	0.8	2.1	1.8	2.6
Dairy	No	18	0.8	2.5	0.8	1.7
	Yes	21	1.0	2.1	0.7	1.4
Poultry	No	45	0.9	3.7	5.3	3.8
	Yes	75	0.5	3.1	3.2	2.0
	Deep pit	76	0.8	2.5	3.0	1.6
Liquid Manure						
Swine	Liquid pit	4	7.2	10.8	7.5	6.6
	Lagoon	1	4.8	6.0	3.6	4.8
Beef	Liquid pit	9	1.3	3.9	2.4	3.5
	Lagoon	1	2.4	4.8	3.6	4.8
Dairy	Liquid pit	8	1.0	4.6	2.2	2.8
	Lagoon	1	2.4	4.8	3.6	4.8
Poultry	Liquid pit	10	1.6	7.2	5.4	3.6

Adapted from: Midwest Plan Service, 1993

Like biosolids, fresh and composted manures contain both organic and inorganic nitrogen, with the highest percentage of nitrogen being organic. Accordingly, it is important to know the mineralization rate to determine how much nitrogen will be available during the growing season. For manure applications, farmers traditionally have relied on tabular guidance provided by land grant universities and other institutions. Table 2-8 shows estimated nitrogen mineralization rates for various manures from one source.

Table 2-8
Nitrogen Mineralization Rates in Manures

Manure Type	Manure Handling	First Year Mineralization Rate (% of Organic N)
Swine	Fresh	50
	Anaerobic liquid	35
	Aerobic liquid	30
Beef	Solid without bedding	35
	Solid with bedding	25
	Anaerobic liquid	30
	Aerobic liquid	25
Dairy	Solid without bedding	35
	Solid with bedding	25
	Anaerobic liquid	30
	Aerobic liquid	25
Poultry	Deep pit	60
	Solid with litter	60
	Solid without litter	60

Adapted From: Midwest Plan Service, 1993

Studies of phosphorus availability are few. In comparing P availability from mineral fertilizers and poultry litter leachate, Sharpley and Sisak (1997) found consistently greater P availability for soils treated with P fertilizer. Other researchers found that WEP of triple-superphosphate (two samples), dairy manure (five samples), and poultry manure (eight samples) was 85%, 48%, and 21%, respectively (Brandt, Elliot, and O'Conner, 2001). By comparison, the researchers found that the WEP of digested biosolids (18 samples), which represent the majority of biosolids produced in the United States, was about 3%. Mineral fertilizers generally have higher WEPs as they are formulated to provide readily available nutrients.

2.2.4 Mineral Fertilizers

Table 2-9 lists some of the myriad mineral fertilizers and their approximate nutrient contents.

Table 2-9
Chemical Fertilizer Nutrient Content (%)

Material	Nitrogen	Phosphate	Potash	Calcium	Magnesium	Sulfur	Chlorine	Copper	Manganese	Zinc	Boron
Nitrogen											
Ammonia,	82	--	--	--	--	--	--	--	--	--	--
Ammonia, Aqua	16-25	--	--	--	--	--	--	--	--	--	--
Ammonium nitrate	33.5	--	--	--	--	--	--	--	--	0.01	--
Ammonium nitrate-limestone mixtures	20.5	--	--	7.3	4.4	0.4	0.4	--	--	--	--
Ammonium Sulfate	21	--	--	0.3	--	23.7	0.5	0.3	--	0.1	--
Ammonium sulfate-	26	--	--	--	--	15.1	--	--	--	--	--
Calcium cyanamide	21	--	--	38.5	0.06	0.3	0.2	0.02	0.04	--	--
Calcium nitrate	15	--	--	19.4	1.5	0.02	0.2	--	--	--	--
Nitrogen solutions	21-49	--	--	--	--	--	--	--	--	--	--
Sodium nitrate	16	--	0.2	0.1	0.05	0.07	0.4	0.07	--	--	0.01
Urea	46	--	--	--	--	--	--	--	--	--	--
Urea-form	38	--	--	--	--	--	--	--	--	--	--
Phosphate											
Basic slag, Open	--	8-12*	--	29	3.4	0.3	--	--	2.2	--	--
Bone meal	2-4.5	22-28**	0.2	20-25	0.4	0.1	0.2	--	--	0.02	--
Phosphoric acid	--	52-60	--	--	--	--	--	--	--	--	--
Rock phosphate	--	***	--	33.2	0.2	0.3	0.1	--	0.03	--	--
Superphosphate,	--	18-20	0.2	20.4	0.2	11.9	0.3	--	--	0.01	--
Superphosphate,	--	42-50	0.4	13.6	0.3	1.4	--	0.01	0.01	--	0.01
Superphosphoric	--	69-76	--	--	--	--	--	--	--	--	--
Potash											
Potassium chloride	--	--	60-62	0.1	0.1	--	47	--	--	--	0.03
Potassium	--	--	22	--	11.2	22.7	1.5	--	--	--	--
Potassium sulfate	--	--	50	0.7	1.2	17.6	2.1	0.001	--	--	0.00
Multiple Nutrient											
Ammoniated super-	3-6	18-20	--	17.2	--	12	--	--	--	--	--
Ammonium	27	15	--	--	--	--	--	--	--	--	--
Ammonium	13-16	20-39	0.2	0.3	0.1	15.4	0.1	0.02	0.2	0.02	0.03
Diammonium	16-21	48-53	--	--	--	--	--	--	--	--	--
Monoammonium	11	48	0.2	1.1	0.3	2.2	0.1	0.02	0.03	0.03	0.02
Nitric phosphates	14-22	10-22	--	8-10	0.1	0.2-3.6	1-12	0.02	0.2	0.02	0.03
Nitrate of soda-	15	--	14	--	--	--	0.5	--	--	--	0.13
Potassium nitrate	13	--	44	0.6	0.4	0.2	1.1	--	--	--	0.1
Wood ashes	--	1.8	5.5	23.3	2.2	0.4	0.2	0.12	0.76	0.2	0.16
Blast furnace slag	--	1.7	0.6	29.3	3.8	1.4	--	--	1.02	0.00	0.01
Dolomite	--	--	--	21.5	11.4	0.3	--	0.001	0.11	--	0.01
Gypsum	--	--	0.5	22.5	0.4	16.8	0.3	--	--	--	--
Kieserite (emjeo)	--	--	--	1.6	18.2	--	--	--	--	--	--
Limestone	--	--	0.3	31.7	3.4	0.1	--	0.004	0.48	0.05	0.00
Lime-sulfur solution	--	--	--	6.7	--	23.8	--	--	--	--	--
Magnesium sulfate	--	--	--	2.2	10.5	14	0.4	--	--	--	--
Sulfur	--	--	--	--	--	30-99.6	--	--	--	--	--

Source: TFI, 1982, Reprinted from the Fertilizer Handbook with the permission of The Fertilizer Institute

As noted above, mineral fertilizer nutrients are readily available for plant use. Exceptions to this are fertilizers that have been treated to slow nitrogen release.

2.3 Metals and Metal Availability

Biosolids producers, regulators, and other parties that deal with biosolids are accustomed to the linkage between metals and biosolids. As evidenced by the federal 503 rule and regulations in many states, total metals content has historically been a measure of biosolid quality. Total metals content is not as commonly associated with non-biosolid soil amendments and fertilizers, however, and less information is available on these products.

In addition to total metals content, the availability of biosolids metals (for plant uptake or leaching into groundwater) has been extensively researched. The literature shows that metals availability has not been assessed as extensively for manures or mineral fertilizers, however.

Metals availability depends upon many factors, one of which is the metal's form. Metals in organic products can exist in many forms. Some of these forms are listed below *in a generally descending order of leachability*.

1. *Water soluble:* metal ions are not tightly bound to soil or organic particles;
2. *Exchangeable:* metal ions are loosely bound to soil or organic particles;
3. *Carbonate-bound:* metal ions are bound to carbonate (if present);
4. *Organically-bound:* metal ions are constituents of or complexed by organic matter in the soil and/or soil amendment; and
5. *Oxide-bound:* metal ions are tightly bound to iron, aluminum, or manganese oxides.

As indicated above, water soluble and exchangeable metal forms exhibit the highest leaching potential. Other forms are less readily available, and can only be made available for leaching by more aggressive means than the addition of water.

Both total metals content and the *relative* availability of metals in soil amendments and fertilizers are presented in the following text. It is critical to remember that the relative availability of different metal forms can only partially reflect the true bioavailability of a metal – many factors impact bioavailability, such as pH, synergistic effects between metals, especially cadmium and zinc, and so forth. Chapter 4 discusses these factors.

2.3.1 Biosolids

Biosolids that are applied to the land must, at a minimum, meet metals limits mandated by the federal 503 regulations and, in most states, limits that are as stringent or more stringent than the 503 limits. Consequently, the concentrations of regulated metals are routinely measured in biosolids. Table 2-10 presents typical biosolids metal concentrations for 503 metals and compares them to the most stringent limits in the regulation. As the table shows, average metal concentrations in U.S. biosolids fall far below the most stringent USEPA criteria.

Table 2-10
Average Biosolids Metals Concentrations (mg/dry kg)

Metal	*Class A Alkaline Stabilized*	*Alkaline and Digested Liquid*	*Alkaline and Digested Cake*	*Composted*	*Heat-dried*	*40 CFR 503, Table 3 Limits*
No. of Wastewater Treatment Plants in Sample	7	48	120	10	5	NA
Arsenic	5.20	5.00	5.00	5.00	5.20	41
Cadmium	2.00	4.05	4.85	4.50	2.00	39
Chromium	38.4	42.2	68.8	52.6	38.4	NA
Copper	122	421	460	380	122	1500
Lead	70.0	56.5	84.6	72.6	70.0	300
Mercury	0.50	2.30	1.70	2.02	0.50	17
Molybdenum	7.50	11.0	12.9	12.9	7.50	75
Nickel	23.0	30.0	36.0	28.1	23.0	420
Selenium	1.70	3.00	4.40	3.90	1.70	100
Zinc	250	752	770	821	250	2800

Source: Lue-Hing et al., 1998

NA = not applicable

The various forms of metals present in biosolids have not been exhaustively studied, but one study (Camp Dresser & McKee Inc. [CDM], 1998) compared metal forms for several biosolids products: a compost, an alkaline stabilized product, and a heat-dried product. Figure 2-2 summarizes the results of that study, which focused on six metals: barium, cadmium, chromium, copper, nickel, lead and zinc.

As Figure 2-2 shows, virtually all metals in the compost and heat-dried product are in organically bound or less-available forms. For the alkaline-stabilized product, 81%, 60%, and 32%, respectively, of the metals barium, copper and nickel in the product are present in the more available leachable and exchangeable forms. (With respect to potential impacts, however, it should also be noted that the total concentration of each of these metals was substantially lower in the alkaline product than in the compost or heat-dried product.)

Figure 2-2
Metal Fractions in Biosolids (CDM, 1998)

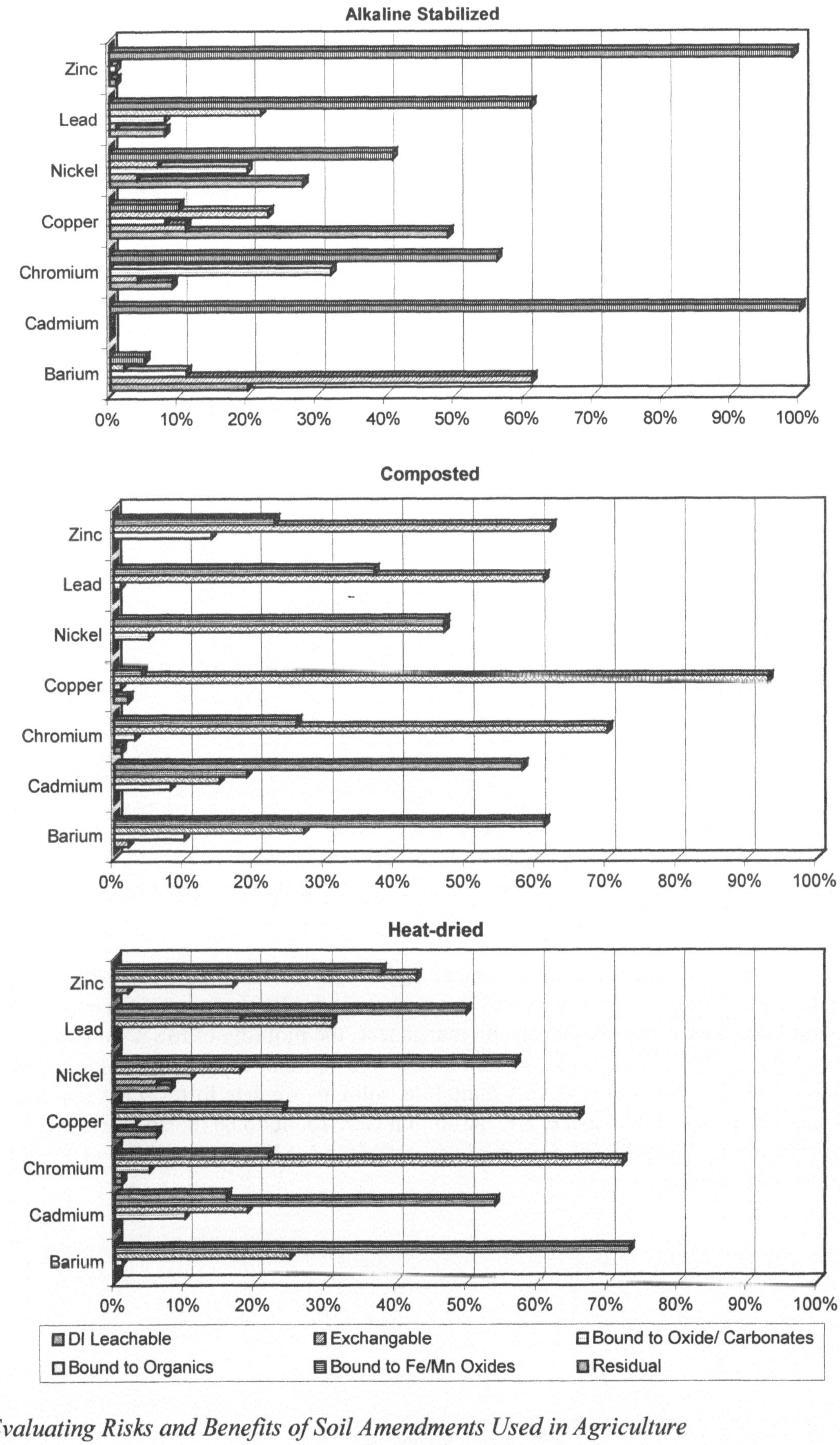

The investigation of metal forms can provide some insight into metals availability, but many researchers use field studies to more directly assess this phenomenon. Chapter 4 discusses these field studies.

2.3.2 MSW and Yard Waste Composts

There have been several major papers written about the heavy metal content of MSW compost (Epstein et al., 1992; Logan et al., 1999). Table 2-11 summarizes the mean heavy metal concentrations found in MSW compost. MSW compost generally complies with the most stringent 503 limits for biosolids, with the occasional exception of lead and arsenic (He, Logan, and Traina, 1995; Logan, et al. 1999).

Table 2-11
MSW Compost Metal Concentrations (mg/dry kg)

Metal	*Mixed MSW Compost (US)*	*Separated MSW Composts (EU)*
Arsenic	2.6	Not Available
Cadmium	2.9	1.1
Chromium	34	29.4
Copper	162	57
Lead	221	112
Mercury	1.2	0.9
Nickel	28	19.9
Zinc	469	281

Adapted from: Epstein et al., 1992

Yard waste metal content has not been extensively investigated, but Golueke, Diaz, and Gurkewitz (1989) reported it to be low. Epstein (1997) presented a range of metal concentrations for raw yard waste (Table 2-12), supporting previous findings that metal concentrations in composted materials should be low.

Very little has been written on metal availability in MSW or yard waste compost, although He, Logan and Traina (1995) explored this issue for ten MSW composts. The researchers concluded that, except for boron and manganese, the mobility of MSW compost metals is expected to be low. They found that some metals (cadmium, copper, lead and manganese) were mainly present in relatively immobile, solid, particulate forms. Zinc was found to be organically complexed, while nickel and chromium were found to be mostly in relatively immobile residual forms. A significant portion of several of the MSW composts' boron content was found to be water soluble.

Table 2-12

Yard Waste Metal Concentrations (mg/dry kg

Metal	*Reported Range*	*Mean*
Arsenic	1.0-12.8	4.8
Cadmium	0.04-0.81	0.32
Chromium	3.7-236	39.4
Copper	8-327	64.0
Lead	11.4-235	69.6
Mercury	0.04-0.5	0.19
Molybdenum	0.05-1.09	0.22
Nickel	3.27-152	26.9
Selenium	0.1-0.55	0.33
Zinc	41.6-295	153

Adapted from: Epstein, 1997

2.3.3 Manures

The metal content of manures has not been thoroughly evaluated, although Sims (1995) reported that metals concentrations in manures might be similar to some biosolids (with respect to 503 pollutants). Table 2-13 supports Sims's report. As data in the table show, the concentrations of some metals are comparable to biosolids, and in some cases, such as arsenic and copper, manures can have higher metal concentrations than biosolids. Metals provided through feed supplements are responsible in part for manure metal content.

With respect to metals (and other chemicals) in manures, information on fate and transport is limited. Sims (1995) notes that:

> "Limited research is available that characterizes the form of these elements [trace, pesticides, organics, hormones] and compounds in animal wastes and even less on any subsequent transformations that might occur in waste-amended soils that could affect their bioavailability and transport."

2.3.4 Mineral Fertilizers

Several recent studies have provided some insight into the metal content of mineral fertilizers. Raven and Loeppert (1997) compared metal concentrations of 35 elements in 24 materials. The researchers found that trace element concentrations generally decreased in the following order: rock phosphate> biosolids> phosphorus fertilizer>other organic amendments and liming materials> potassium fertilizers>nitrogen fertilizers.

USEPA (1999c) recently surveyed the metal content of primary nutrient, secondary nutrient, and micronutrient fertilizers (1999c). Table 2-14 presents that study's results. As the table shows, the metal content of primary nutrient fertilizers is generally lower than that of zinc and other micronutrient fertilizers, which have relatively high concentrations of multiple heavy metals. The bioavailability of metals in fertilizers has been investigated to a small extent through field studies (see Chapter 4).

Table 2-13
Comparison of Manure and Biosolids Metal Concentrations (mg/dry kg)

Metal	Dairy Cattle Manure (1)	Beef Cattle Manure(1)	Swine Manure(2)	Poultry Manure(3)	Biosolids Median(4)
Arsenic	NA	NA	3.7	13	5.0
Cadmium	0.25	NA	2.5	2.4	4.4
Copper	38	36	109-501	465	425
Lead	NA	NA	7.6	46	76
Mercury	NA	NA	ND	NA	1.8
Molybdenum	6.2	4.94	2.6	19	12
Nickel	23	NA	29	16	33
Zinc	150	129	455-656	602	735

ND = not detected; NA = not available

(1) Adapted from ASAE, 2000

(2) Compiled from ASAE, 2000; Epstein, 1999

(3) Compiled from ASAE, 2000; Texas Agricultural Extension Service (2001)

(4) Average from data in Lue-Hing et al. (1998), weighted by number of samples

Table 2-14
Mean Heavy Metal Concentrations in Mineral Fertilizers (mg/kg)

Product	*Arsenic*	*Cadmium*	*Chromium*	*Copper*	*Lead*	*Mercury*	*Nickel*	*Zinc*
Phosphate	11.3	65	173	56.6	12.2	0.1	27.5	240
NPK for Phosphorus	12.8	30.6	83.9	31.4	217	0.07	28.5	234
NPK for Nitrogen	5.3	5.0	50.6	41.3	31.8	0.2	11.8	205
Potash	0.3	NA	0.5	1.0	NA	0.1	1.4	1.6
Sulfur (nutrient)	5.0	36.6	37.9	24.3	5.4	0.1	48.9	385
Sulfur (pH adjustment)	7.8	NA	NA	40.4	6.4	NA	NA	NA
Lime	14.6	2.5	21.2	42.0	44.5	0.1	7.9	353
Gypsum	4.5	1.7	1.4	23.7	4.6	0.011	3.0	53.8
Iron	1,662	177	NA	667	7,135	NA	NA	NA
Boron	531	0.75	1.3	8.1	5.5	0.2	2.5	6.0
Manganese	7.8	2.3	10.0	11.3	27.5	0.01	50	60.8
Zinc	20	398	339	1,101	10,013	3.36	890	NA
Micronutrient Mixes	41.9	27.9	230	29,650	1,798	0.1	12.5	77,300

NA = not available
From: USEPA, 1999c

2.3.5 Product Comparisons

Based upon information presented earlier in this chapter, Table 2-15 compares the metal concentrations of biosolids, manures, and selected mineral fertilizers. As shown in Table 2-15, the concentrations of some metals in these products can be comparable.

It is critical to remember, however, that mineral fertilizers and organic soil amendments are applied to the land at dramatically different rates, and any comparison of their potential impact must account for this difference. Table 2-15 also illustrates the importance of product application rate. To develop the figure, we assumed that organic amendments were applied to corn at the agronomic rate for nitrogen, superphosphate was applied to meet the phosphate requirements of corn, and a micronutrient mix was applied to meet primarily zinc needs. The above approach is not a rigorous comparison of the metal application rates, but it does reflect the critical point that the quantities of micronutrient fertilizers and, to a lesser extent, primary nutrient fertilizers, are small compared to amendments used for fertilization. Accordingly, the quantity of metals applied with mineral products can be smaller on a per site basis than amendment-applied metals, although the metal concentrations in these products (especially micronutrients) can be higher. Cadmium supplied through phosphate applications is an exception to this generalization.

The comparison of application rates also illustrates another key point: metals applied through manure or manure-based products can exceed metals applied through biosolids applications. Our example indicates that arsenic and copper from poultry manure applications and nickel from swine manure applications fall into this category. The impact of manure-applied metals has not been intensively investigated to our knowledge, but based upon the above information, such an investigation may be warranted.

2.4 Biological Characteristics

Because mineral fertilizers are, by definition, inorganic, they are not discussed in this chapter. Soil amendments are organic, however, and can contain beneficial — as well as potentially harmful — microorganisms. Generally, the literature indicates that researchers have been focusing more on pathogens than on beneficial organisms in biosolids or manures.

2.4.1 Biosolids

Microorganisms that may impact health are generally combined under the term "pathogens," which may include bacteria, viruses, and parasites. Table 2-16 presents some of the pathogens that may be present in untreated sewage sludge, and the potential health consequences of each.

Because it would be impractical to monitor each of the pathogens in Table 2-16, *indicator organisms* are often used to determine biosolids quality with respect to pathogens. The premise behind this approach is that similar organisms (for example, bacteria) will react similarly to environmental pressures. In other words, a process that destroys one type of indicator pathogen should correspondingly reduce the population of similar pathogens. Presently, fecal coliform and *Salmonella* are primary indicator organisms upon which the 503 pathogen reduction requirements are based (see Chapter 1). Some pathogens are more resistant than others, however, and for that reason, the 503 rule also includes criteria for enteric viruses and helminth ova.

According to 503 requirements, most wastewater treatment plants monitor biosolids for fecal coliforms and *Salmonella.* Table 2-17 summarizes the biosolids pathogen indicator content compiled by Lue-Hing et al. (1998) for AMSA and compares them to 503 requirements.

Table 2-15
Estimated Land-Applied Metals Quantities per Hectare

	Metal Concentration (Average), mg/dry kg				
Metal	***Biosolids***	***Swine Manure***	***Poultry Manure***	***Phosphate Fertilizer***	***Micronutrient Mix***
Arsenic	5.01	3.7	13	11.3	41.9
Cadmium	4.45	2.5	2.4	65	27.9
Chromium	59.3	NA	NA	173	230
Copper	425	305	465	56.6	29,650
Lead	75.9	7.6	46	12.2	1798
Mercury	1.79	ND	NA	0.1	0.1
Molybdenum	12.1	2.6	19	NA	NA
Nickel	33.2	29	16	27.5	12.5
Selenium	3.8	NA	NA	NA	NA
Zinc	735	555	602	240.2	77,300
	Applied Metals (g/ha)				
Metal	***Biosolids*** (1)	***Swine Manure*** (2)	***Poultry Manure*** (3)	***Phosphate Fertilizer*** (4)	***Micronutrient Mix*** (5)
Arsenic	31.3	26.7	80	5.38	1.17
Cadmium	27.8	18.1	15	30.9	0.78
Chromium	371	NA	NA	82.3	6.44
Copper	2,654	2,204	2,854	26.9	830
Lead	475	54.9	282.3	5.81	50.3
Mercury	11.2	NA	NA	0.0	0.003
Molybdenum	75.5	18.4	NA	NA	NA
Nickel	208	212	NA	13.1	0.35
Selenium	24.1	NA	NA	NA	NA
Zinc	4,596	4,010	3,694	114	2,164

Notes:

(1) Applied to meet N requirements of corn, 2.15% N available in first year, application rate of 6.25 dry MT/ha, Metal concentrations from Table 2-13 (chromium and selenium averaged from Lue-Hing, et al. (1998))

(2) Applied to meet N requirements of corn, 1.86% N available in first year, application rate of 7.23 dry MT/ha, Metal concentrations from Table 2-13

(3) Applied to meet N requirements of corn, 2.19% N available in first year, application rate of 6.14 dry MT/ha Metal concentrations from Table 2-13

(4) Applied to meet P requirements of corn, application rate of 476 kg/ha of a product with 20% P_2O_5 content Metal concentrations from Table 2-14

(5) Applied at rate of 2.24 kg zinc/ha, application rate of 28 kg micronutrient mix/ha. Metal concentrations from Table 2-14

Table 2-16
Potential Pathogens in Untreated Sewage Sludge

Pathogens	Disease
Bacteria	
Shigella sp.	Bacillary dysentary
Salmonella sp.	Salmonellosis (gastroenteritis)
Salmonella typhi	Typhoid fever
Vibrio cholerae	Cholera
Escherichia coli	Gastroenteric diseases
Yersina sp.	Yersinosis (gastroenteritis)
Campylobacter jejuni	Campylobacterosis (gastroenteritis)
Viruses	
Hepatitus A bisrus	Infectious hepatitus
Norwalk viruses	Acute gastoenteritis
Rotaviruses	Acute gastoenteritis
Polioviruses	Poliomyelitis
Coxackie viruses	"flu-like" symptoms
Echoviruses	"flu-like" symptoms
Protozoa	
Entamoeba histolytica	Amebiasis (amoebic dysentary)
Giardia lamblia	Giardiasis (gastroenteritis)
Cryptosporidium sp.	Cryptosporidiosis (gastroenteritis)
Balantidium coli	Balantidiasis (gastroenteritis)
Helminths	
Ascaris sp.	Ascariasis (roundworm infection)
Taenia sp.	Taeniasis (tapeworm infection)
Necator americanus	Ancylostomiasis (hookworm infection
Trichuris trichuria	Trichuriasis (whipworm infection)

Source: NRC, 1996

Recently, contamination of water supplies by several emerging pathogens has led to questions regarding the concentration of these pathogens in biosolids. Pathogens that fall into this category include *Escherichia coli* O157:H7 (which can cause illness (diarrhea) and, in some cases, death in humans (Valcour et al., 2002), *Giardia* and *Cryptosporidium*. Methods to easily measure these parameters at low concentrations, however, are not yet fully developed, although research is underway to develop analytical methods.

Table 2-17
Biosolids Pathogen Contents

Pathogen	*Class A Alkaline Stabilized*	*Class A Composted*	*Class A Heat-dried*	*40 CFR 503 Class A Standards*
No. of Plants in Sample	5	4	5	NA
Fecal Coliform (MPN/g Mean)	3	76	6	<1,000
Fecal Coliform (MPN/g Median)	1	506	8	NA
Salmonella sp (MPN/4g Mean)	2	2	0	<3
Salmonella sp (MPN/4g Median)	2	2	0	NA
Pathogen	***Class B Alkaline Stabilized***	***Class B Digested Liquid***	***Class B Digested Cake***	***40 CFR 503 Class B Standards***
No. of Plants in Sample	55	26	41	NA
Fecal Coliform (MPN/g Mean)	60	104,600	6,521	<2,000,000
Fecal Coliform (MPN/g Median)	9600	472,600	16,071	NA
Salmonella sp (MPN/4g Mean)	2070	NA	NA	NA
Salmonella sp (MPN/4g Median)	4000	NA	NA	NA

Source: Lue-Hing et al., 1998

NA = not applicable

In addition to pathogen content at the completion of processing, pathogen regrowth has been raised as an issue for some Class A products. Class A processes significantly eliminate pathogens and, since parasites and viruses cannot multiply outside of a host, the only concern for pathogenic regrowth relates to bacteria. There has been some evidence that the indicator organism *E. coli* may regrow if reintroduced into compost (Soares et al., 1985), although the potential for regrowth depends on moisture content and compost stability; in a study of 16 Massachusetts biosolids composts, 3 exhibited pathogen regrowth. Evaluating *Salmonella*, Yanko (1988) also found that finished compost product may have significant increases in bacterial population, even though few salmonellae were detected in the compost coming out of the active composting phase. These studies have also found that the *Salmonella* population regrowth is transient and that there is subsequent die off (Epstein and Donovan, 1992). In a separate study, Hussong, Burge and Enkiri (1985) found that competition from native microbes is the most important factor in maintaining low *Salmonella* populations in finished compost.

2.4.2 MSW and Yard Waste Composts

There is very little information on pathogens in MSW and yard waste compost. For MSW compost, Golueke, Diaz, and Gurkewitz (1989) reported that *E. coli* was found in concentrations above 1.1 x 10^8, with fecal and total coliforms reaching concentrations greater than 1.1 X 10^8 and 4 x 10^8, respectively. They did not detect S*almonella.* With respect to yard waste compost, the authors reported that "*pathogens of most concern either were present in low concentrations or could not be detected.*" Meckes et al. (1995) found yard waste compost fecal coliforms ranging from <2.5 most probable number per gram (MPN/g) to 2, 300 MPN/g, with an average of 38 MPN/g. They did not detect *E. coli.* The authors noted that there is potential for contamination of yard waste composting from machinery, wastewater treatment facilities, manure, etc., and that because yard waste composting is not done according to 503 requirements, pathogens are not necessarily killed off. In general, pathogen concentrations are well below USEPA limits, but the potential still exists for pathogenic growth.

2.4.3 Manures

The predominant human pathogens found in manure are *Salmonella* sp., *Escherichia coli* O157:H7, *Cryptosporidium parvum*, *Listeria monocytogenes*, *Mycobacterium paratuberculosis*, and *Giardia* sp. (Pell, 1997; Olson et al., 1999). Table 2-18 shows the organisms and the diseases that they can transmit.

As with biosolids, the pathogen content of manures is sometimes assessed through indicator organisms. For example, Sims and Wolf (1994) noted that bacterial indicators such as fecal coliforms or *E. coli* are used to assess potential water resource contamination, due to the difficulty and expense in conducting specific pathogen analysis in poultry wastes.

Manures may contain large numbers of bacteria. For example, fecal coliform concentrations in fresh manures can range from 0.5 x 10^7 to more than 3 x 10^{10} colonies/dry g in dairy, beef, swine, and some poultry manures (ASAE, 2000). Jones (1980) reported that *salmonella* in English cattle generally are on the order of 100 cells/gram, although 3.1 x 10^3 cells/ gram feces were found in infected dairy cows (Jones et al., 1982). Reductions in these levels might be expected due to the elapsed time between excretion and land application, as well as various levels of treatment (i.e, anaerobic lagoons, storage/stacking) that are employed at most livestock operations (Kintzer, 2002). Nonetheless, fecal coliforms and *Salmonella* concentrations in biosolids, especially Class A products, are lower, as shown in Table 2-17. Several studies have shown that *E. coli* was present in feces of cattle, especially calves (Zhao et al., 1995; Hancock et al., 1994). Recently, attention has focused on a particular pathogenic strain of *E. coli— E. coli* O157:H7. Cattle and other ruminants appear to be primary carriers of this pathogen (Goss, et al. 2002), although it can also be found in other animals (deer, rodents, and pet dogs, for example). The researchers note that this pathogen is detected infrequently in swine and poultry manure, however. Reported concentrations of *E. coli* O157:H7 in cattle manures range from at 100 to 10,000 colony forming units/g (fresh weight) (Goss et al., 2002; Zhao, et al., 1995).

Some parasites have also been noted in manures. Calves can excrete *Cryptosporidium* oocysts in large numbers. Up to 10^{10} oocysts /g feces have been reported (Olson et al., 1997; O'Handley et al., 1999). Calves can also pass up to 100,000 *Giardia* cysts per gram of feces (Olson et al., 1997; O'Handley et al., 1999). Studies have shown that about 20% of dairy and beef calves tested positive for *Cryptosporidium* (Pell, 1997).

Table 2-18
Some Microorganisms in Animal Manures that Can Cause Disease in Humans

Bacteria	Disease
Campylobacter jejuni	Bloody diarrhea, abdominal pain
Bacillus anthracis	Skin disease
Brucella abortus	Gastrointestinal disease; anorexia
Escherichia coli	Gastrointestinal disease
Leptospira spp.	Kidney infection
Listeria monocytogenes	Meningitis in adults; infection in infants
Mycobacterium tuberculosis	Tuberculosis
Mycobacterium paratuberculosis	Johnes disease
Salmonella spp.	Salmonellosis
Yersinia enterocolitica	Gastrointestinal infection
Viruses	**Disease**
Adenoviruses	Eye and respiratory infections
Avian enteroviruses	Respiratory infections
Avian reoviruses	Infectious bronchitis
Bovine parovirus	Respiratory infection
Bovine rhinovirus	Foot and mouth disease
Enterovirus	Respiratory infection
Reovirus	Respiratory infection
Rhinovirus	Parainfluenza
Rotaviruses	Gastrointestinal infections
Protozoan Parasites	**Disease**
Balatidium coli	Blantidias
Cryptosporidium parvum	Cryptosporidiasis
Eimeria spp.	Coccidiosis
Giardia lamblia	Giardiasis
Toxoplasma spp.	Toxoplasmosis
Helminth Parasites	**Disease**
Ascaris lumbricoides	Ascariasis

In two recent studies, the Centers for Disease Control and Prevention (CDC) analyzed indicator organisms in poultry and swine manures, and *Cryptospridium* for swine only (Karpati and Rubin, 1998; Campagnolo and Rubin, 1998). Surprisingly, the poultry study found *E. coli* in only 40% of the manure samples. The studies attributed the low detection frequency to sample collection or analysis methods, although other factors (age of manure) could potentially have played a role as well. No bacteria concentrations were provided in the poultry study, as its major focus was antiobiotics in manures and bacterial resistance (these issues are discussed in Chapter 2.5.3). The CDC study of swine manure did, however, report bacteria and parasites concentrations, which are summarized in Table 2-19.

Table 2-19
Pathogens in Lagooned Swine Manure

Farm No.	*Salmonella* sp. (MPN/100mL)	*E.coli* (MPN/100mL)	*Enterrococcus* sp. (MPN/100mL)	*Yersina* sp. (MPN/100mL)	*Cryptosporidium parvum* (oocysts/L)
1	740	100,000	60,000	<300	2,250
2	<300	380,000	27,000	<300	NA
3	<300	<100	1,800	<300	1,560
4	<300	270,000	390,000	<300	NA
5	<300	69,000	12,000	<300	NA
7	9,300	310,000	1,900,000	<300	NA
8	<300	140,000	38,000	<300	1,230

Source: Campagnolo and Rubin, 1998

Although types of viruses in animal manures have been reported, no quantitative data was found (Strauch, 1991).

Manure contains numerous beneficial organisms including bacteria and fungi. Little data is available on the beneficial use of these organisms, but a review of the literature on beneficial organisms found in poultry manure was reported by Sims and Wolf (1994). They reported total microbial viable colony forming units (CFU) counts of 10^8 to 10^{10} per gram dry waste and fungal populations of up to 2.6 x 10^5/g dry waste.

2.5 Other Chemical Characteristics

In addition to nutrients and metals, soil amendments and fertilizers may contain minute quantities of other chemicals that are introduced through raw materials or, in the case of amendments, medicines. Potential chemicals that may be included in amendments and fertilizers are discussed below.

2.5.1 Biosolids

Biosolids contain a variety of chemicals including organics, pesticides and herbicides, but the concentrations of detected pollutants have fallen dramatically over the last 20 years. Table 2-20 compares detected organics from a 40-plant study in the 1970s to the concentrations of those same pollutants from a 180-plant survey in the late 1980s (the "National Sewage Sludge Survey," or NSSS (USEPA, 1999b)). Industrial pretreatment programs, designed to limit the introduction of pollutants into the sewer system, are largely credited with the improvement in biosolids quality (NRC, 1996).

Today, concerns regarding biosolids quality focus on dioxin, endocrine disruptors (natural and synthetic) and pharmaceuticals.

Table 2-20
Trends in Biosolids Organic Compound Content (1)

Organic Pollutants	Percent Detection 1970's	1980's	Mean Values(2) 1970's	1980's
Aldrin	16	3	6.4	1.9
Benzene	93	0	1,784	--
Benzo(a)pyrene	21	3	138	--
Bis(2-ethylhexyl)phthalate	100	62	155 mg/kg	74.7 mg/kg
Chlorodane	16	0	6.4	--
Dieldrin	16	4	6.4	--
Heptachlor	16	0	6.4	--
Hexachlorbenzene	16	0	155	--
Hexachlorobutadiene	5	0	23	--
Dimethylnitrosamine	5	0	6.4	--
Lindane	16	0	57	--
PCBs	NA	NA	--	--
Toxaphene	16	0	6.4	--
Trichloroethene	84	1	8,139	--
DDD/DDE/DDT	NA	NA	--	--

Adapted from: NRC, 1996
(1) Data from 40 Cities study conducted in late 1970's and National Sewage Sludge Survey (NSSS) conducted in late 1980's.
2) All units in microgram/dry kilogram (ug/kg) unless otherwise noted.

Concerns regarding *dioxin*, a name used to describe any of a family of compounds known as dibenzo-p-dioxins, however, have led to a dioxin limit in proposed revisions of the 503 rule. Some dioxin forms are more toxic than others, and, thus, levels of dioxin forms (or cogeners) are usually referred to by their toxic equivalent (TEQ). (The total dioxin TEQ is essentially a sum of the concentration of each dioxin form multiplied by its individual Toxic Equivalent Factor, or TEF.) In its proposed regulation, USEPA has set a risk-based limit of 300 parts per trillion (ppt) toxic equivalents for dioxin. However, most biosolids in this country have much lower levels of this ubiquitous compound. A survey by the Association of Metropolitan Sewerage Agencies (AMSA) found biosolids dioxin concentrations to be quite low, averaging 48.5 ppt-TEQ (Alvarado, et al., 2001). Median dioxin TEQs in European biosolids are reportedly somewhat higher, ranging from 21 to 90 ppt (Jones and Sewart, 1997). USEPA notes that dioxin levels in biosolids have been relatively constant over time and may even be decreasing (Landers, 2000).

Other ubiquitous chemicals perceived to be potential concerns in land-applied biosolids include brominated diphenyl ethers (BDEs, used in flame retardants), surfactants (including alkylphenols and linear alkylbenzene sulfonates, LAS), and estrogens. Hale et al. (2000) report concentrations of 1,100 to 2,290 microgram/dry kg for "penta-like" BDEs and a wider range of 84.6 to 4,890 microgram/dry kg for deca-BDE's. Dentel et al. (1993) reports that LAS concentrations of between 0.5% and 1.5% (dry weight basis) have measured in biosolids, particularly anaerobically digested materials (these compound are readily degraded by aerobic processes such as aerobic wastewater treatment, but are not removed by anaerobic processes). Biosolids also contain estrogens (Shore, Correll, and Chakaroborty, 1995). Although data is limited, Shore (1993) reports that water extracts of biosolids contain from 18 to 70 nanograms per liter (ng/L) of estrogen and 10 to 174 ng/L testosterone. Busheé, Edwards, and Moore (1998) found 5.2 micrograms per dry kg of the natural estrogen 17β-estradiol in a municipal biosolid (compared to a concentration of 35.1 micrograms per dry kg in a horse manure/bedding sample). Both surfactants and estrogens are considered to be "endocrine disruptors" because of their potential impacts on hormonal systems in humans and wildlife (this issue is discussed further in Chapter 4).

Pharmaceuticals, such as antibiotics, may potentially be present in biosolids as well, as 30% to 90% of medicines taken by humans (and animals) are excreted in urine as active substances (WEAO, 2001). We are unaware of any studies, however, quantifying pharmaceuticals in biosolids.

2.5.2 MSW and Yard Waste Compost

Primary concerns with respect to MSW and yard waste composts include organic contaminants and herbicides, respectively. Dioxins have also been identified in these composts.

Because MSW may be a mix of plastics, paints, and other discarded chemicals, it potentially could contain a wide range of organic contaminants including acetone, toluene, dichloromethane, trichloroethylene, xylene. Many organic compounds are likely to break down or volatilize during composting, but certain *xenobiotics* (chemicals that are not produced by living organisms) are more difficult to decompose and may have carcinogenic potential. Table 2-21 lists these compounds (Logan et al., 1999).

Two herbicides — clopyralid and picloram — have been found in some yard waste composts in Washington, Pennsylvania, and New Zealand (Bezdicek et al., 2001). The impact of yard waste compost contamination with these herbicides and remedial actions taken are described in Chapter 4.

Epstein (1997) notes that dioxins have been found in a variety of composts, and noted a TEQ of 7.7-41 ppt in yard waste compost.

2.5.3 Manures

Concerns have been raised regarding the concentrations (and impact) of endocrine disruptors, antibiotics, growth hormones and pesticides in manures.

Limited data suggest that manures contain a significant amount of naturally occurring estrogens (Goss et al. 2002), which fall into the category of endocrine disruptors. Estradiol, estrone, and equol are some of the estrogens that the researchers (Goss, et al., 2002) identified in manures through a literature survey on this subject. The researchers also indicate that the estrogen detected in the highest concentration in manures (equol) is much less potent than another estrogen — estradiol-17β — which is also present.

Table 2-21
Xenobiotics that Could Potentially Contaminate MSW Compost

Compound	Use or Source
Alkylated polychlorinated dibenzofurans	Chemical by-product and combustion
Penta- to nonylphenols	Degradation of polyethoxylate surfacants
Phthalate esters	Plasticizers
Polybrominated dibenzo-p-dioxins	Degradation of polyethoxylate surfacants
Phthalate esters	Plasticizers
Polybrominated biphenyls	Fire retardant
Polybrominated dibenzo-p-dioxins	Chemical by-product and combustion of dibenzofurans
Polychlorinated biphenylenes	Combustion of PCBs
Polychlorinated diphenyl ethers	Chemical by-product and combustion
Polychlorinated hydroxydiphenyl ethers	Chemical by-product and combustion
Polychlorinated napthalenes	PCB replacement
Polychlorinated quaterphenyls	Formed from PCBs
Polychlorinated terphenyls	PCB replacement
Tetrachlorobenzyltoluenes	PCB replacement

From: Logan et al., 1999, Reprinted from Compost Science and Utilization with permission through the Copyright Clearance Center

As Table 2-22 shows, the CDC found antibiotic residues in many samples of lagooned manure (one antibiotic was found in a monitoring well near the sample site as well). Both the CDC swine manure study and a separate study of poultry manure (Karpati and Rubin, 1998) found evidence of antibiotic resistance in *E. coli*, *Salmonella* sp, and *Enterococcus* sp. detected in study samples.

Karpati and Rubin (1998) concluded that the study by CDC could not confirm the source of the resistance (the animal's digestive system or the environment), but that the presence of antimicrobial resistance in the environment warranted further investigation.

Table 2-22
Antibiotics in Lagooned Swine Manure

Farm No.	Tetracycline (ug/L)	Sulfonamide (ug/L)	β-Lactam (ug/L)	Macrolide (ug/L)
1	250	>20	>2	227
2	11	>20	>2	<10
3	150	>20	>2	60
4	68	>20	3.5	<10
5	66	>20	2.1	81
7	540	>20	2.1	275
8	11	>20	2.9	15

Source: Campagnolo and Rubin, 1998

The CDC found small concentrations of pesticides in fields and water samples around the swine production site it studied, but it did not detect pesticides in manure samples (Campagnolo and Rubin, 1998).

2.5.4 Mineral Fertilizers

In addition to metals, chemicals of potential concern that have been found in some mineral fertilizers include dioxin and, potentially, perchlorate.

In 1998, the Washington State Department of Ecology sampled 50 fertilizers (including 19 agricultural fertilizers). It found that most of these fertilizers had low or undetectable concentrations of dioxins, but several had high concentrations, as well (WSDE, 1999). An agricultural micronutrient, believed to be derived from steel mill flue dust, had a TEQ of 140 ppt, and a home-use fertilizer had a TEQ of 54 ppt. A sampling in 1997 revealed another micronutrient derived from steel mill flue dust to have a TEQ of 340 ppt (WSDE, 1999). Table 2-23 shows the results of dioxin analyses for agricultural fertilizers.

A contested USEPA survey of fertilizers in 1999 reportedly found perchlorate – a compound that can impact thyroid function (EPA, 2001b) - in 9 fertilizer samples and 8 horticultural fertilizer components (Susarle, et al., 1999), but subsequent research indicates that it may be present only in fertilizers derived from Chilean caliche (USEPA, 2001c), a natural perchlorate source. Concentrations in Chilean nitrate salts ranged from 1440 microgram/gram (ug/g) to 1700 ug/g when independently tested in 5 different laboratories. Chilean nitrate salts constitute only about 0.14% of the fertilizer applied in the United States and, though insufficient information exists with which to assess risk, the potential for exposure to perchlorate from fertilizers appears to be small (USEPA, 2001c).

2.6 Odor Potential

Traditionally, odor studies of biosolids and manures have focused on biosolids processing facilities, livestock operations, biosolids or manure products, and odor mitigation measures. Here in the United States, however, it appears that little work has been done to measure odor or gaseous emissions at land application sites for manures (MEQB, 1999) or biosolids. Some studies of ammonia emissions from manure application sites have been performed (Zupanci et al., 1998; Byers and Mikkelsen, 1999). (Ammonia, however, is not a great contributor to offsite odors at great distances from land application sites — unlike hydrogen sulfide, organic sulfurs and other "pervasive" compounds found in biosolids and manures that can be detected far from their point of origin).

Nonetheless, the land application of these materials is acknowledged to be a significant source of odor complaints (Ministry of Agriculture, Fisheries and Food, MAFF,1998) and public opposition (National Biosolids Partnership, NBP, 2001). In the UK, a 1995/1996 survey found that 64% of the complaints regarding swine operations involved land application: for poultry and cattle operations, land application accounted for 40% and 74% of the complaints received, respectively (MAFF, 1998). The findings that land application is the single greatest source of

Table 2-23
Dioxins in Agricultural MIneral Fertilizers

Bulk/Packaged Fertilizers	Sample Year	TEQ (ppt)*
Allied Minerals Dolomite	1997	Not detected
Fort James Nutri Lime	1997	35
Fort James Nutri Lime	1998	7.4
Holnam Cement Kiln Dust	1997	0.95
Northwest Alloys High Mag-Gro	1998	0.01
Unocal Ammonium Nitrate	1998	0.0
Agrium Ammonium Phosphate Sulfate	1998	0.0
IMC Kalium Potash	1998	0.02
Kelly Green Recycled Fresh Fish	1998	1.4
UAP 0-45-0	1998	0.05
High Yield Sulfur	1998	0.0
Micronutrients	**Sample Year**	**TEQ (ppt)***
Bay Zinc Tire Dust	1997	1.6
Bay Zinc Blu-Min	1997	340
Bay Zinc LHM	1997	5.6
Bay Zinc Liquid	1997	0.64
Frit F-503G Sample #1	1998	27
Frit F-503G Sample #2	1998	140
RSA Ruffin-Ready Zn	1998	<0.01
Tech-Flo Zeta Zinc 22	1998	0.01
Green Label Super Starter	1998	0.29
Cozinco Sample #1	1998	0.00
Cozinco Sample #2	1998	0.07
Bioplus Micro 700	1998	0.00
Horizon Ag Micro-Plus	1998	<0.01
Western Farm Monteray 9% Zinc	1998	<0.01
Monteray 10% Zinc	1998	0.04
Hydro-Agri/Viking Ship FS/31	1998	0.00

Source: Washington State Department of Ecology, 1999
*ppt = parts per trillion.
Solid samples on a dry weight basis. Liquid samples on a volume basis.
Dioxin cogeners not detected are set at zero.

complaints from livestock operations confirms the results of an earlier UK study (Hardwick, 1985). With respect to U.S. operations, Jacobson et al. (1998) note that observations in the Midwest and Minnesota seem to agree with this finding. The authors further note, however, that the increased use of manure injection/incorporation and longer manure storage times (which can both be odor mitigation measures) in this country may result in an increased percentage of odor complaints originating from manure storage and livestock operations themselves. To our knowledge, no similar assessments of biosolids land application odors have been performed.

Generally, many of the factors that impact odor emissions from manure applications are the same as those impacting biosolids applications (MAFF, 1998). Accordingly, some odor mitigation measures may be appropriate for both (although the relative benefits observed may depend upon the initial odor of the product and other factors). For example, the injection of liquid biosolids into the soil has long been known to reduce application odors. Studies with liquid manures (Phillips et al., 1990) indicate that deep injection, shallow injection and band spreading reduced application odors by 83%, 70% and 38%, respectively, compared to conventional broadcasting. Others have noted similar and significant odor reductions (MAFF, 1998) for these approaches as compared to surface spreading techniques.

Another approach to reduce land application odors is to reduce the odor of the land-applied material itself. For biosolids, a variety of stabilization techniques (see Chapter 1) have been used to reduce odor (as well as significantly reduce pathogens). Stabilization of manures is less prevalent, but appears to be gaining popularity. At least one study (Pain et al., 1990) found that the stabilization of swine manure with anaerobic digestion reduced land application odors by 80%. Research continues on a variety of alternative odor-reduction measures, including dietary manipulation, chemical additives (to the manure) and microbiological additives.

Although odors have historically been viewed as a nuisance issue, some now question whether they can impact public health as well. This issue is explored in greater depth in Chapter 4 of this document.

2.7 Summary

As this chapter has shown, the characteristics of biosolids, MSW, and yard waste composts, manures and mineral fertilizers vary considerably. Nonetheless, there are some simple generalizations that can be made:

- ***Nutrients:*** The nutrient content of soil amendments is generally far lower than mineral fertilizers and, accordingly, amendments must be applied at much higher rates to meet plant nutrient requirements (or be supplemented by mineral fertilizers).
- ***Metals:*** Metal concentrations in manures and, particularly, in micronutrient fertilizers can exceed biosolids metals concentrations. As Table 2-15 shows, the higher concentrations in fertilizers are generally offset by the small amount of these concentrated materials that must be applied. Manure application rates are similar to application rates for biosolids, however, and therefore, application rates for some metals can be similar for the two products.
- ***Pathogens:*** Indicator pathogen concentrations in manures can exceed corresponding concentrations in biosolids, especially Class A biosolids. The impact of this higher pathogen content on public health is explored in Chapter 4.

CHAPTER 3.0

PRODUCT BENEFITS

Despite common knowledge that soil amendments can transform poor soils into productive soils, negative perceptions about biosolids and manures continue to overshadow their value in sustainable agriculture. Conversely, the benefits of mineral fertilizers with respect to crop yields tend to be more clearly understood because fertilizers are manufactured specifically to supply the nutrients needed to grow plants. This chapter attempts to overcome some of the negative perceptions regarding soil amendments by clarifying their benefits. It clearly defines the benefits of soil amendment use in agriculture in terms of improved soil properties; crop yields and reduced crop production costs; carbon sequestration; and disease suppression. It also discusses the benefits of mineral fertilizers.

3.1 Amendments and Soil Quality

Traditionally, the value of soil amendments has been determined by crop yield and the market value of macronutrients, and occasionally, micronutrients. (USDA, 1998). Unfortunately, this approach does not consider a more critical role that they play: *improving soil quality.*

The importance of soil quality cannot be underestimated and soil amendments can aid in improving soil quality. As noted by USDA (1997):

> Maintaining and improving the quality of the Nation's soils can increase farm productivity, minimize use of nutrients and pesticides, improve water and air quality, and help store greenhouse gases.

Unfortunately, however, multiple factors work to degrade soil quality, including erosion; excess tillage, soil compaction by farm machinery, salinization, biological degradation, and acidification.

Erosion is perhaps one of the most prevalent degraders of soil worldwide, carrying away valuable topsoils and reducing farm productivity. Excess tillage can also remove topsoil and its organic matter, leaving the soils vulnerable to erosion and nutrient depletion. Soil compaction by farm machinery can make tilling difficult, hinder seedling emergence from the soil and root growth, and increase water erosion (USDA, 1997). Soils low in organic matter are particularly susceptible to compaction. Soil compaction's economic impact is profound: losses from compaction in Minnesota, Wisconsin, Iowa, Illinois, Indiana, and Ohio were estimated in 1990 to be as high as $100 million per year [Erdat and Voorhees (1990), as reported in USDA, 1997]. Salinity is another concern and a problem that affects an estimated 48 million acres of agricultural land (USDA, 1997). Finally, biological degradation is perhaps the most significant concern, because "it affects the life of the soil, and because organic matter significantly affects the physical and chemical properties of soils" (NRC, 1993).

From the above factors, it is clear that the loss of organic matter is an essential threat to long-term soil fertility (Draeger et al. 1999). Thus, preventing the loss of organic matter is a key to maintaining healthy soil.

Soil amendments can restore soil organic matter and improve soil quality. Overall, their long-term application can result in superior soil physical condition compared to conventionally fertilized and unfertilized plots (Wood and Hattey, 1995). Evaluating manures, Stewart, Robinson & Parker (2000) further note that the impact of soil amendments can be long-lasting. They note that approved soil tilth and fertility were observed in fields where manures had been applied 20 years earlier. Conversely, fields treated with mineral fertilizer show few beneficial effects unless treatments were continuous.

The following subchapters highlight researchers' understandings of the specific benefits of soil amendment use.

3.1.1 Soil Organic Content

Many of the benefits associated with soil amendments are attributable to the organic matter that they provide. As noted by WSDE (2000), soil organic matter "buffers the soil from extremes of the surrounding environment (e.g., drought, nutrient supply, contamination), ultimately reducing risk for the farmer. Increasing the content of soil organic matter can improve crop production, and rebuild the soil's 'bank account' for the future."

Specifically, increased organic matter can improve physical properties (water retention, soil structure, water infiltration, bulk density, porosity), chemical properties (Cation Exchange Capacity, pH), and in some cases, biological properties. Most of these changes have been noted with applications of non-composted biosolids (Oberle and Keeney, 1994; Lindsay and Logan, 1998), compost (Rynk, 1992; Gouin, 1997; Tester, 1990) and manure (CAST, 1996).

Long-term studies have revealed that the application of biosolids can dramatically increase soil organic matter. Over a 10-year period of biosolids applications by the Metro Wastewater Reclamation District (Denver, Colorado), Jacobs et al. (1993) showed a two-fold increase in soil organic matter from 0.6% to 1.24% in one field (Field 88). Another field (Field 89) showed an increase from 0.43% to 1.24%. For both fields, the greatest increase in organic matter was realized upon the highest biosolids application of the study period (11.5 dry ton /acre (dt/ac) or 25.8 dry metric/hectare (mt/ha) for Field 88 and 18 dt/ac or 40.3 mt/ha for Field 89). Over a 10-year period from 1975 to 1985, analyses of a soil receiving biosolids from the

Metropolitan Water Reclamation District of Greater Chicago showed that the organic carbon increased from 0.95% to 5.06% (Jacobs et al., 1993).

Researchers have also found that composts benefit soil organic content. Bevacqua and Mellano (1994) found that the addition of biosolids compost to soil over a 2-year period increased organic matter from 0.77% to 1.50%.

In a comprehensive survey, Wood and Hattey (1995) found that that organic carbon from manure accumulates in surface soils with repeated manure applications. However, the manure-derived buildup of organic matter is dependent on manure type, application rate, and soil texture. For example, Johnston, et al. (1989), found that soil organic matter from swine manure in a sandy loam and clay loam increased over an 11-year period about 15% and 44%, respectively. However, organic matter did not increase when the same amount of manure (495 ton/ac or 1109 Mg/ha over 11 years) was applied to a silt loam.

Ndayegamie and Côté (1989) found that solid cattle manure and liquid swine manure applied to a silty loam (4% organic content) both increased soil organic carbon, but that the increases were greater for the cattle manure. They further noted that the impact of manure applications on organic matter was far more pronounced at higher application rates (e.g., 27 ton/ac or 60 Mg/ha), a finding that other works support.

Other studies have also revealed that soil organic content increases linearly with biosolids' application rate (Lindsay and Logan, 1998; Stowe and Caudill, 1994). Figure 3-1 illustrates carbon increases resulting from a single application of an anaerobically digested biosolids. The figure shows that carbon increases proportionally to application rate, and that a single application of organic matter (in this case, biosolids applied in 1990 only) remains in the soil, although it can decompose somewhat over time.

Tester (1990) also noted that some organic matter from soil amendments decomposes or leaches in each cropping season (about 24% and 37% of the annually added organic matter for a biosolids compost and beef manure, respectively). Figure 3-2 illustrates this phenomenon.

Organic matter decomposition can explain why appliers may not see a large organic matter increase after a single small application and why it may take several years to realize the benefits of organic matter addition (Jacobs et al., 1993; Wood and Hattey, 1995).

Numerous factors, including soil type, influence the speed at which soil organic matter degrades, and, correspondingly, the benefits that are realized. For example, a short-term study with liquid cattle manure found that higher application rates were needed to maintain soil organic matter (at pre-study levels) in a sandy loam than in a sandy clay or a clay (Spallacci and Boshi, 1985). The researchers attributed the higher application requirements of sandy loam to better aeration and, accordingly, a faster organic matter degradation rate. Soil organic matter is reduced by aerobic microbial activity, which degrades organic matter. Other factors that can affect the decomposition rate are: amendment composition (especially, total carbon content and carbon-to-nitrogen (C:N) ratio); soil temperature and moisture; and method of application (Lindsay and Logan, 1998).

Figure 3-1
Impact of Biosolids Application Rate on Soil Organic Matter

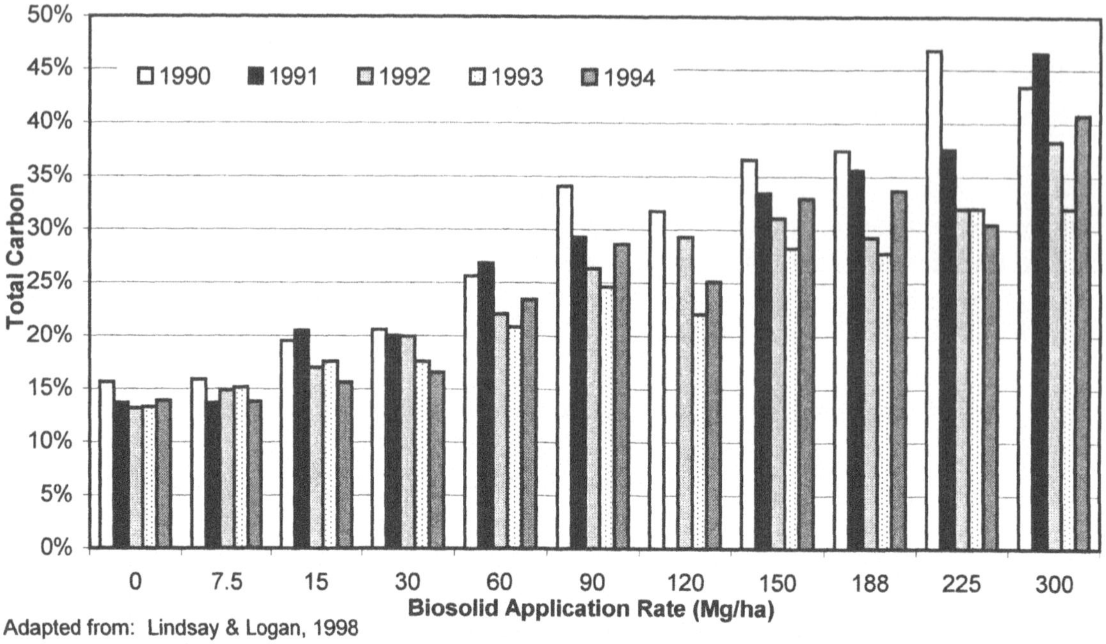

Adapted from: Lindsay & Logan, 1998

Reprinted from the Journal of Environmental Quality with permission through the Copyright Clearance Center

3.1.2 Soil Structure

Organic matter improves soil structure by binding soil particles (soil aggregation) and creating large pores through which air and water can move (i.e., increasing soil porosity). Organic amendments are also less dense than soils, so they can reduce soil bulk density (NRC, 1996). Increased porosity and decreased density can also reduce penetration resistance in soils. Both biosolids and manures have been shown to reduce penetration resistance in a sandy soil (Tester, 1990).

Figure 3-2
Impact of Manure & Composted Biosolids on Soil Organic Carbon

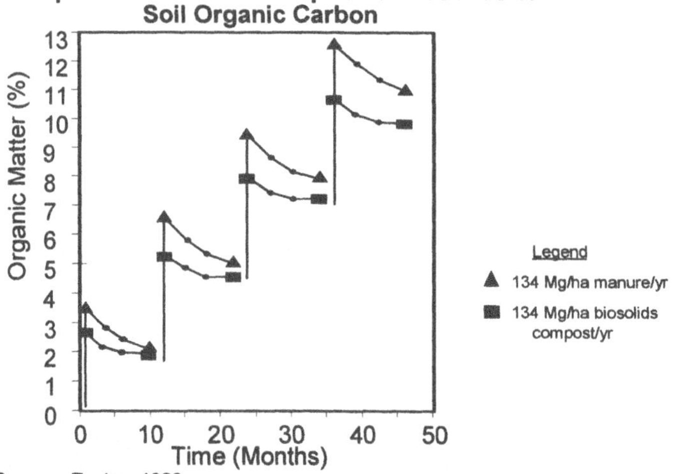

Source: Tester, 1990

Reprinted from Soil Science of America Journal with permission from the Soil Science Society of America

Studies have demonstrated that biosolids applications decrease soil bulk density (Lindsay and Logan, 1998; Khaleel, Reddy, and Overcash, 1981; Zebarth et al., 1999). As Figure 3-3 shows, Lindsay and Logan have observed that soil bulk density decreased with an increase in biosolids application rate.

Figure 3-3
Impact of Biosolids on Soil Porosity and Bulk Density

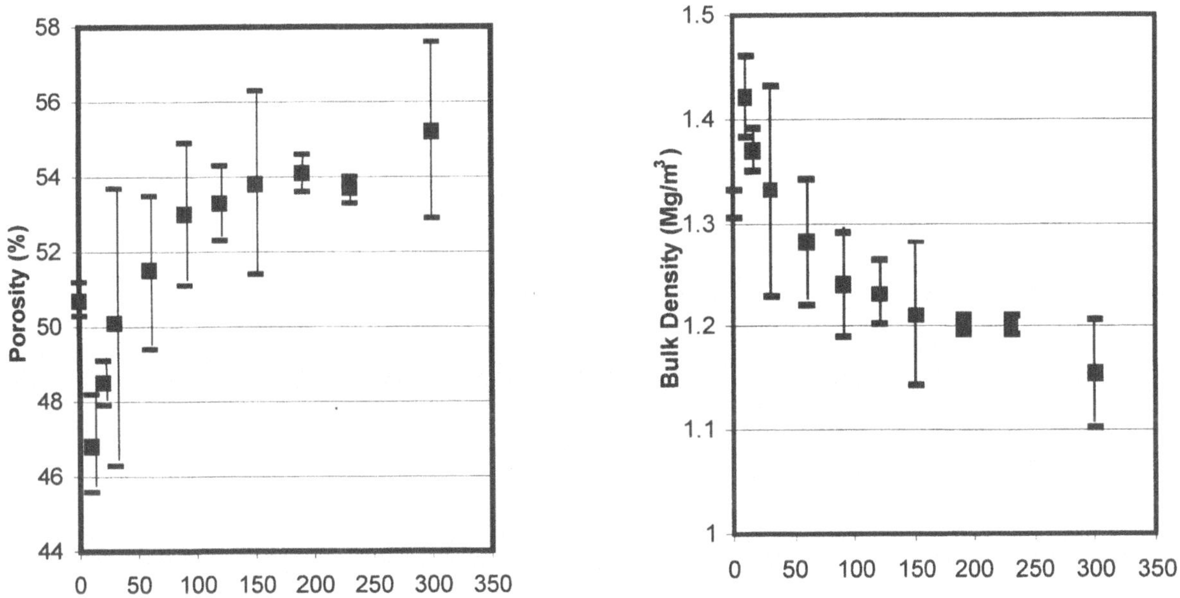

From: Lindsay & Logan, 1998. Reprinted from the Journal of Environmental Quality with permission through the Copyright Clearance Center.

Giusquiani et al. (1995) noted a linear relationship between increases in soil porosity and the amount of MSW compost added to soil. They observed that the number of elongated pores, in particular, which are important to root growth, drainage, and soil aeration, had increased as more compost was added to the soil. The increased pore space had resulted in lower bulk density and better water retention. Turner et al. (1994) also measured decreased bulk density due to MSW compost amendment although, over time, the compost began to break down, and subsequently, bulk densities increased.

Others have noted a direct correlation between compost applications and bulk density. North Carolina State Agricultural Extension studies found that when a soil was blended with 25% MSW compost, bulk density decreased proportionately to the amount of compost added to the soil (Shelton, 1991).

In a comprehensive survey Wood and Hattey (1995) found that manures lower soil bulk density. Specifically, they cite a 90-year study in Denmark (Schonning, Christensen and Carstensen, 1994) that compared soil characteristics of sandy loam planted over a 4-year rotation with winter wheat, beets or turnips, spring barley, and a clover/grass mixture. In this study, fields were either not fertilized, fertilized with inorganic fertilizer, or amended with manure. The manure-amended fields had the highest organic content in surface soils (0–20 cm or 0–8 in), and had lower bulk density than unfertilized or mineral-fertilized plots.

By reducing soil bulk density and increasing soil porosity, amendments also reduce the soil's resistance to root penetration. Tester (1990) showed that, even 5 years after the amendments were applied, fields treated with biosolids, composts, and manures had a lower surface soil penetration resistance than unamended or conventionally fertilized fields. Figure 3-4 illustrates this phenomenon for composted biosolids.

Figure 3-4
Impact of Composted Biosolids on Soil Penetration Resistance

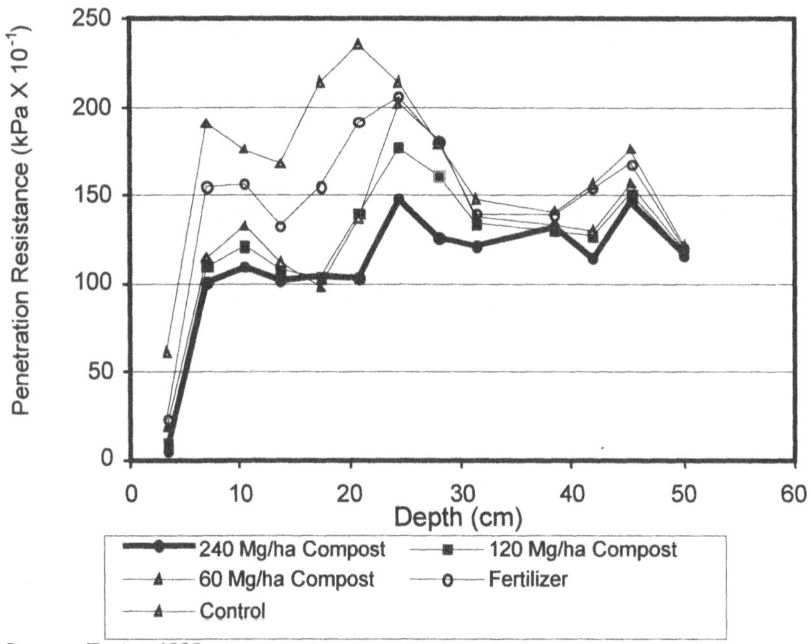

Source: Tester, 1990
Reprinted from Soil Science of America Journal with permission from the Soil Science Society of America

3.1.3 Water Retention

Soil amendments can beneficially affect water retention (the length of time that a soil can hold water), infiltration, and in some cases, water holding capacity (how much water a soil can hold).

A multiyear study in Yuma, Arizona (Jacobs et al., 1993) showed that biosolids additions reduced irrigation requirements by 26% (compared to fields that had never received biosolids), while increasing alfalfa and barley yields. The site manager noted, however, that reduced water consumption and increased crop yields were not observed after a single application but were observed only after repeated applications.

Bali et al. (1998) reported that biosolids application rates ranging from 4 to 12 ton/ac (9 to 27 Mg/ha) before the first irrigation of the season significantly improved the infiltration rate of some California heavy clay soils. Water infiltration improvements increased with an increase in application rates.

Shiralipour et al. (1996) found that the addition of co-composted biosolids and yard waste increased the soil's water holding capacity by 5% to 15%, depending on soil type.

Zebarth et al. (1999) studied a variety of organic materials including biosolids compost. They found that the compost increased the soil's ability to retain water.

Manures have also been shown to increase water infiltration rates (CAST, 1996).

3.1.4 Soil Cation Exchange Capacity

Cation exchange capacity (CEC) is a measure of a soil's ability to retain cations (positively charged ions). Organic matter has negative charges that help hold positively charged nutrients such as ammonium, calcium, and others in the soil (Reeves, Thornberry and Carey, 1998).

Laboratory studies (Epstein, 1975) have shown that the organic matter provided by the addition of biosolids can increase a soil's CEC and, hence, its ability to retain nutrients; field studies have supported these findings. A WERF survey of long-term biosolids application programs (Jacobs et al. 1993) found several examples of CEC increases from biosolids applications. The Denver Metro Wastewater Reclamation District land application operation showed that soil CEC more than doubled (from 6.0 to 14.6 meq/100g in Field 88 and from 5.0 to 14.3 meq/100g in Field 89) after biosolids applications of 11.5 dt/ac (25.7 mt/ha) and 18 dt/ac (40.3 mt/ha), respectively. The land application of a Chicago biosolids at an average rate of 27.5 dt/ac (61.6 dry mt/ha) increased soil CEC from 13.4 meq/100g to 27.4 meq/100g over a 10-year period.

North Carolina State Agricultural Extension studies (Shelton, 1991) found that clay and sand soils blended with 25% MSW compost increased soil CEC by 500% to 600%.

With respect to manures, a detailed literature review by Wood and Hattey (1995) indicated that soil CEC in manure-amended soils remained constant or increased linearly with manure type and application rate.

3.1.5 Soil Nutrient Restoration

Crops remove nutrients from the soil, some more than others. The use of manures and biosolids, however, can restore soil nitrogen, phosphorus, sulfur and other nutrients. Nitrogen is usually the most limiting nutrient for plant growth, however, and has accordingly been a research focus.

As noted earlier, much of the nitrogen in soil amendments is present in organic form and is released slowly to the environment through mineralization. Accordingly, nitrogen can slowly accumulate in the soil with repeated applications of soil amendments. This phenomenon was demonstrated through a study at the Rothamsted Experiment Station in England that spanned more than a century. Researchers found that, after 140 years, the total N content of manure-amended soil was 2.25 times greater than the N content of unamended soils or soils treated with inorganic fertilizer, which remained essentially unchanged over this same period (Johnston, et al., 1989). The researchers also observed that the increase of N had slowed over time, mirroring an effect also found by shorter-term studies (e.g., Sommerfeldt, Chang and Entz, 1988, who found that N accumulations gradually decrease after 20 to 30 years of manure applications).

As Chapter 2 shows, both biosolids and manures also contain micronutrients that can correct deficiencies in soils. For example, cattle feedlot manure and biosolids applications have been shown to correct iron deficiencies in sorghum (Mathers et al., 1980; McCaslin et al., 1987). In Colorado, the copper provided by biosolids applications was greatly appreciated by farmers, as their soils are copper deficient — in fact, the farmers were not pleased when copper levels in the biosolids that they were applying fell from the 1,500 to 2,000 ppm range to 850 ppm (Scharp, 2000). Additionally, one biosolids user in the area noted that he gets better results with biosolids than mineral fertilizers because of the micronutrients supplied (he does not typically apply commercial micronutrient mixes because of cost).

3.1.6 Soil Microbes

The organic material that is supplied by soil amendments can increase microbial biomass and help maintain beneficial bacterial and fungi populations (Humpert, 2000), as well as stimulate microbial activity (CAST, 1996).

Long-term addition of organics can alter the structure and function of soil microbes (Wood and Hattey, 1995). Ndayegamie and Côte (1989) found that bacteria, actinomycetes, fungi, and potentially mineralizable N increased (as compared to control) over an 8-year application of liquid swine manure and solid cattle manure. In this study, higher actinomycetes and fungal populations were observed for the cattle manure (as compared to the swine manure). In a series of laboratory and field experiments, Lazarovits and Conn (1997) also found that soil microbial populations increased with manure applications, but that responses were more favorable to some manures than others - all amendments studied (poultry litter, solid cattle manure and liquid swine manure) generally increased soil microbial populations by orders of magnitude (with the exception of one laboratory experiment with swine manure), but poultry litter appeared to effect the largest increase. For example, in one experiment, poultry litter increased the soil total bacterial population from 10 million colony forming units (CFU)/g soil to 6 billion CFU/g soil, and the total fungal population from 100,000 CFU/g soil to 3 million CFU/g soil. The researchers attribute higher increases from poultry litter to a higher nitrogen content. Lazarovits and Conn (1997) also note that the effects are seasonal: after 21 weeks, microbial populations had generally returned to pre-application levels.

3.1.7 Erosion Reduction

Several studies have been found that note the erosion-reduction benefits of land-applied biosolids (Smith, 1996; Draeger et al., 1999; Oberle and Keeney, 1994) and other soil amendments (Humpert, 2000).

Research appears to focus on areas with severe erosion control needs, such as non-vegetated or newly vegetated highway embankments and land reclamation projects. Typically, higher amendment application rates are required to reduce erosion in these situations, and results obtained may not be as dramatic at the lower application rates typically used in agriculture. Nonetheless, the successful use of soil amendments in these extreme cases illustrates the erosion reduction potential of these materials.

Sort and Alcaniz (1996) found that high biosolids application rates reduced erosion on non-vegetated, amended soils more than 90% compared to un-amended, non-vegetated soil. Where vegetation was well developed, erosion was also lower where biosolids were applied. In a simulated rainfall event, soil loss decreased with increasing biosolids application rate, and when biosolids were applied directly on the soil surface, erosion rates and particle mobilization caused by raindrop impact were minimal. The researchers also found that biosolids amendments increased infiltration rates and improved soil structure, which decreases both runoff and erosion.

Extensive work with composted dairy manure has also illustrated the beneficial effects of soil amendments on erosion (McCoy and Cogburn, 2001). Through a series of demonstrations on highway embankments across the state of Texas, the Texas Natural Resource and Conservation Commission (TNRCC) has successfully vegetated highway slopes using composted manure to control erosion. In a 1999 demonstration, the use of compost resulted in thick grass growing within 2 months of the compost application, although five previous attempts to vegetate the slope using other means had failed.

3.2 Amendment Impacts on Crop Yields and Production Costs

Crop yields corresponding to organic matter addition are usually improved, but exceptions may occur due to the multiple variables that affect plant growth. These variables include crop type; soil type; climate, soil and crop management practices; and the characteristics of the organic materials that are applied.

Nonetheless, numerous studies have shown that biosolids-, compost-, and manure-amended fields can produce crop yields comparable to, or superior to, mineral-fertilized fields. Some of these studies are highlighted later in this chapter.

Impacts on production costs are harder to assess. Production costs can take into consideration increased crop yield; fertilizer value of organic amendments and reduced fertilizer requirements; reduced water consumption; reduced pesticide use; increased application requirements; and other factors. Traditionally, economic comparisons of soil amendment use have focused on fertilizer replacement value alone because other benefits, such as improved soil health and reduced soil erosion potential, are difficult to quantify.

The subchapters below focus on crop yields and the fertilizer replacement value of amendments because available literature predominately discusses these aspects. The other benefits listed above should not be forgotten, however, as their individual and cumulative benefits may far exceed mere nutrient value.

3.2.1 Biosolids

Research has shown that annual — and even single — applications of biosolids can produce crop yields comparable to, or superior to, mineral fertilizer applications.

Parr and Hornick (1992) found that corn yield from a single biosolids application could actually increase over time and compared favorably with crop yields from fields that were treated annually with a conventional fertilizer (Table 3-1).

Table 3-1
Effect of a Single Biosolids Application on Corn Grain Yields

Treatment	Corn Grain Yields (hundreds of kg/ha) 1972	1973	1974	1975	1976	Average
Biosolids*						
0 dry Mg/ha	25.1	21.1	6.8	16.2	0.91	4.0
56 dry Mg/ha	55.4	68.7	64.6	54.4	38.5	56.3
112 dry Mg/ha	66.5	102.8	68.7	66.0	67.1	74.4
224 dry Mg/ha	61.2	108.4	76.5	69.4	69.2	76.9
NPK Fertilizer**	34.3	51.0	57.6	50.1	47.7	48.2

* Biosolids applied in 1972 only; no inorganic fertilizers were applied.

**Fertilized plots received annual applications of 180 kg/ha N, 40 kg/ha P, and 75 kg/ha K; no biosolids were applied

Source: USDA, 1998

A comprehensive WERF study (Jacobs et al., 1993) found positive crop responses from long-term biosolids application sites across the nation, as the next three case studies indicate. A multi-year study in Yuma, Arizona showed increased crop yields for alfalfa and barley crops that had received biosolids (Figure 3-5) as compared to other crop yields in the county. The site manager estimated that the amount of irrigation water used on biosolids-amended fields is about 26% lower than that used on fields upon which biosolids had not been applied. At an application site in Virginia, soybean, corn, and wheat yields were comparable to other areas in the county over a 3-year period. A farm in North Carolina that produces corn, sorghum, and Bermuda grass on biosolids-amended fields has won numerous local, regional, and state awards for its high corn and sorghum crop yields.

Two independent studies at Midwestern sites found that corn yields from liquid biosolids-amended fields were as good, if not better than, yields from conventionally fertilized fields (Peterson et al., 1994; Clapp, et al.; 1994). A long-term study in Wisconsin showed corn yields in fields that were annually amended with biosolids to be commensurate to those from a field that was treated each year with conventional fertilizer (Peterson et al., 1994), as Figure 3-6 shows. (**Note:** All biosolids plots received a one-time treatment of 0-0-60 fertilizer in 1981 to meet crop potassium needs.) A 16-year study near Rosemount, Minnesota had similar results (Clapp et al., 1994), with corn yields from fields receiving biosolids averaging 156 bushels/acre (bu/ac) (8,600 kg/ha), compared to average yields of 145 bu/ac (7,700 kg/ha) for fields being managed by traditional fertilizer methods. Reed canarygrass yields on study fields where biosolids were applied averaged 5 tons/ac/yr (11.2 Mg/ha/yr), and where traditional fertilizers were applied, averaged 4.3 tons/ac/yr (9.6 Mg/ha/yr). Both studies found that an oversupply of phosphorus (as compared to plant needs) did not upset nutrient balances in crops. Tables 3-2 and 3-3 compare corn grain and canarygrass yields for biosolids-amended and traditionally fertilized fields.

Figure 3-5
Alfalfa Barley Yields (1988-1990): Biosolids Amended Fields vs. County

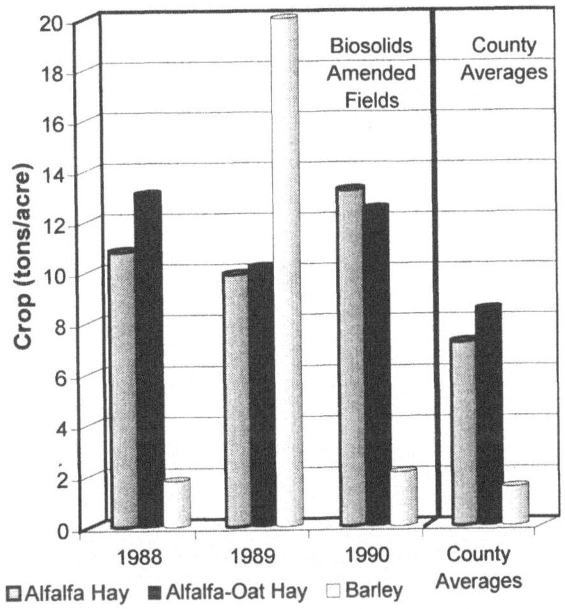

Adapted from Jacobs et al,1993

Figure 3-6
Corn Grain Yields in Elkhorn, WI, from Biosolids Applications

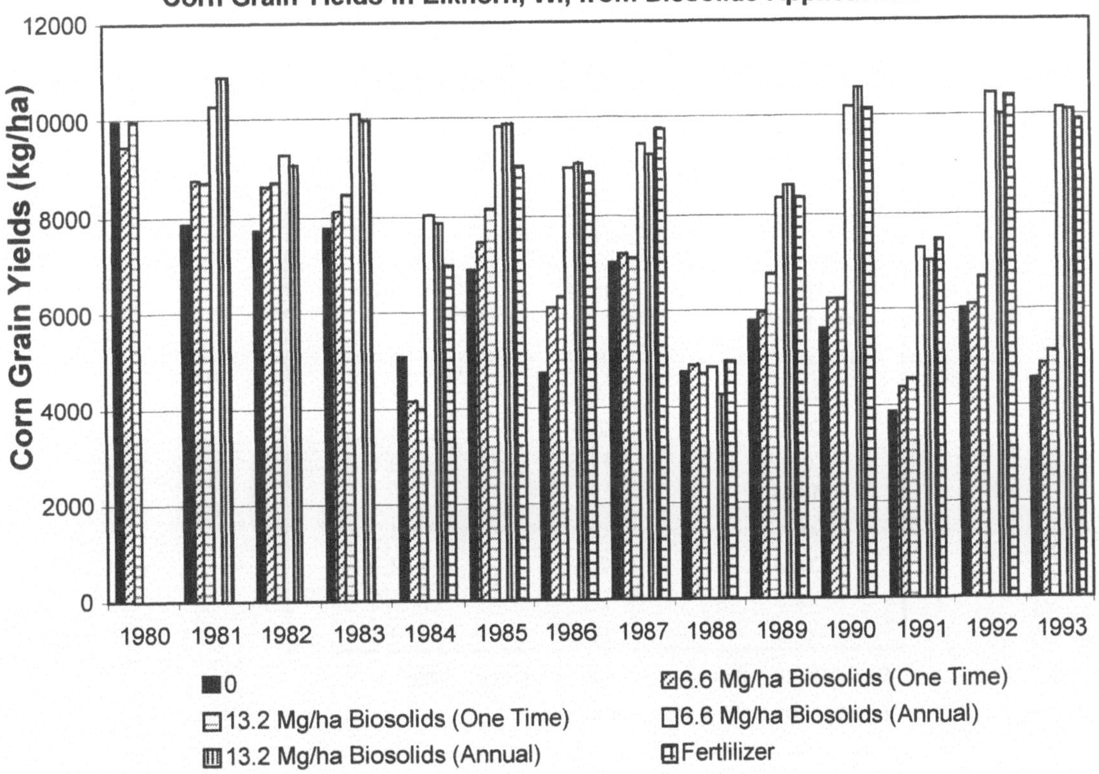

Adapted from: Peterson et al., 1994

Table 3-2
Corn Yields on Biosolids- and Mineral-Fertilized Fields

Year	Units	Grain Yield Biosolids	Grain Yield Fertilizer	Fodder Yield Biosolids	Fodder Yield Fertilizer
1976-1979	Mg/ha	8.8	8.3	18.1	18
	bu/ac	165	156		
1980-1983	Mg/ha	8.1	7.6	19.4	15.5
	bu/ac	152	142		
1984-1987	Mg/ha	8.3	7.4	15.7	14.7
	bu/ac	155	139		
1988-1992	Mg/ha	8.3	7.9	16.7	16.6
	bu/ac	156	149		
Mean	Mg/ha	8.4	7.7	17.4	15.9
	bu/ac	156	145		

Biosolids applied at an average annual rate of 11 Mg/ha

Adapted from: Clapp et al., 1994

Reprinted from Sewage Sludge: Land Utilization and the Environment with permission through the Copyright Clearance Center.

Table 3-3
Reed Canarygrass Yields on Biosolids- and Mineral-Fertilized Fields

Year	Dry Matter Yield (Mg/ha) Biosolids	Fertilizer
1976-1979		
Cut 1	5.2	4.5
Cut 2	3.4	3.1
Cut 3	2.3	1.8
Total	10.9	9.4
1980-1983		
Cut 1	5.6	5.3
Cut 2	3.5	2.6
Cut 3	2.2	1.8
Total	11.3	9.7
Mean		
Cut 1	5.4	4.9
Cut 2	3.4	2.8
Cut 3	2.2	1.8
Total	11.1	9.6

Biosolids applied at an average annual rate of 15 Mg/ha

Adapted from: Clapp et al., 1994

The fertilizer replacement value of biosolids was estimated for two land application programs, one on each coast of the United States. The value of a Virginia biosolids was estimated to be $51.74/acre ($127.75/ha) for primary nutrients alone and $67.52/acre

($166.72/ha) including secondary and micronutrient value (Faust and Oberst, 1996). In Washington state, a total value of $22.58/dry ton ($24.90/dry Mg), or $90.52/acre ($223.51/ha) for a 4 dry ton/acre (8.96 Mg/ha) application was estimated for primary nutrients and sulfur (Sullivan, 1998).

3.2.2 Composts

Each of the above studies focused on noncomposted biosolids, but biosolids and MSW composts have been shown to provide similar benefits. Bevacqua and Mellano (1994) found that onions, snapdragons, turf, and spinach responded positively to biosolids compost additions, with yields increasing by 5%. Shiralipour et al. (1996) found that lettuce grew best on soils amended with 30 to 60 tons of compost per acre (or 67 to 134 mg/ha). Shiralipour et al. (1993) also summarized the results of several studies of compost applications on crop yield, which Table 3-4 presents.

Only one article quantified biosolids compost use in terms of crop production costs. Litvany (1998) estimated the reduction in fertilizer and chemical control costs if compost were used on citrus groves. Based on the conventional application rates of fertilizers, pesticides, herbicides, and lime, Litvany estimated the savings for compost use to be $85 per acre. Citrus production could be increased by 8%, further benefiting farmers.

Table 3-4
Effects of Compost Applications on Crop Growth and Yield

Plant Species	Compost Type(s)	Application Rate (ton/ac)	Crop Response
Corn	MSW, Biosolids	4.5-200	10% to 135% yield increase
Sorghum	MSW	3-150	10% to 25% yield increase
Wheat, Barley, Oats, Rye	MSW, fertilizer	2.5-200	10% to 173% yield increase
Grass and Pastures	MSW, MSW/biosolids	15-58	Increased yields at rates up to 32 tons/ac
Cucumber	MSW/ N-K-Mg fertilizer	6-12	18% to 20% yield increase

Adapted from: Shiralipour, 1993

Others have also found MSW compost applications to be beneficial. North Carolina State Extension Researchers (Shelton, 1991) found that MSW compost amendment increased plant growth, regardless of the amount of mineral fertilizer given to the plants. Tomatoes positively responded to compost with better plant growth and more fruit production. Rodd, et al. (1999), and MacLeod, et al. (1999) found that MSW compost improved the growth of barley, wheat, and potatoes. However, compost was not appropriate for all crops; the pH was too high for acid-loving plants.

Several researchers have studied the benefits of yard waste compost. Maynard (1999) found that 50 ton/ac (112 Mg/ha) applications of yard waste compost can be used in place of one application of 10-10-10 mineral fertilizer. The study also found that tomato growth was enhanced when compost was used in conjunction with fertilizer. Kostewicz (1993) found that yard waste compost that is applied at rates of 25 to 100 tons/ac (56 to 224 Mg/ha) enhanced pole bean growth and yield, but that unstable compost diminished bean yields. Stephens and Kostewicz (1994) found that yard waste compost, if sufficiently aged, could enhance plant growth if used in conjunction with nitrogen supplements.

3.2.3 Manures

CAST (1996) reported many studies that have shown that the land application of manures will produce crop yields equivalent to, or superior to, yields from commercially fertilized fields, as well as crop quality improvements. They note that the reasons for crop responses that were superior to commercially fertilized fields included manure-supplied nutrients and/or improved soil conditions from manure application.

USDA (1998) noted several studies that have cited the benefits of manure applications. They reported that beef cattle manure use resulted in yield increases for corn silage, corn grain, grain sorghum, forage sorghum, and perennial forage crops from beef cattle manure (or manure effluent applications). They also reported rice yield increases of up to 286% with poultry litter additions. These yields were higher than those obtained with inorganic NPK and zinc additions.

As noted earlier, most estimates of soil amendment value have focused on fertilizer replacement value, and the few studies that the project team found addressing the value of manure adopted this approach. In a survey of Wisconsin dairy operations, Nowak, Shepard and Madison (1998) estimated that farmers could save $15.70/acre ($38.77/hectare) on commercial fertilizer through proper use of on-farm nutrient sources. Sutton, et al. (1974) estimated that manure fertilizer value could range from $3.65/ton ($4.02/Mg) for dairy manure with bedding to $36.01/ton ($39.70/Mg) for deep pit poultry manure.

3.3 Carbon Sequestration

With growing concern over global warming, conservationists are seeking ways to reduce carbon dioxide emissions, a contributor to that phenomenon. One approach that agriculture is adopting is carbon sequestration, or carbon dioxide reduction by ensuring that carbon remains in the soil. To sequester carbon, no-till farming and other conservation practices (minimum-till, buffer strips, and reforestation) are some of the specific approaches that have been used. In Iowa, a pioneering program in 1999 offered farmers "carbon credits" for their adoption of conservation practices (Caspers-Simmet, 2000).

Although the use of soil amendments has not been specifically noted in the Iowa program, soil amendment use offers many of the same benefits as traditional conservation practices, and it sequesters carbon, as well. Specifically, soil amendment applications reduce erosion, improve soil productivity and health, and replenish soil carbon through the addition of organics.

Whether farmers using such materials are involved in a carbon credit program or not, their contribution to reducing greenhouse gases should not be ignored.

3.4 Disease Suppression

Researchers have observed disease suppressive benefits for composts and manures. Compost applications have been shown to suppress certain plant diseases, resulting in a decreased reliance on pesticides (Humpert, 2000). Composted leaves and agricultural waste are, in fact, being used at California berry farms and grape vineyards as part of Integrated Pest Management systems (Goldstein, 1998; Farrell, 1996).

Most of the literature on the disease-suppressive nature of compost is based on work conducted with composted mushroom waste, manure, and agricultural wastes. Only Dickerson (1999) used biosolids compost in a study to evaluate disease suppression. This study revealed that the application of 10 to 20 tons/ac (22 to 45 Mg/ha) of biosolids compost reduced the number of chili pepper plants lost to *Phytophthora* root rot. Plants were still susceptible to chile wilt (which is spread by an insect).

The addition of compost to soil suppresses disease by several means: competition between plant pathogens and microbes present in the compost; antibiotic production; predation and parasitism; and systemic resistance by plants as induced by compost (Hoitink et al., 1997). Compost has been found to reduce incidents of *Phytophthora* and *Pythium* root rot, *Rhizoctonia Solani.*

Compost tea, or watery compost extract, has also been studied. Spraying plants with a tea made from composted manure and bedding has been found to suppress Grey mold and damping off (Riggle, 1996; Brinton, Tranker, and Droffner, 1996).

Working with a variety of manures, Lazarovits and Conn (1997) found that manures could reduce plant pathogens responsible for Verticillum wilt (in potatoes) and potato scab. Factors that impacted the disease suppressive effect of manures included soil type, soil moisture and, most importantly, the manure type used. For example, in some soils, the researchers found that fresh poultry litter and liquid swine manure decreased plant pathogens, but that solid cattle manure appeared to have no effect. Composting the poultry litter destroyed its disease suppressive characteristics.

Lazarovits and Conn (1997) note that the use of soil amendments to suppress plant diseases results in a healthier soil than the use of chemical fumigants for this purpose. They describe the soil amendments studied (which also included paper mill sludges and other materials) as selective fumigants, noting that they decrease plant pathogens but increase overall soil microbial populations. In contrast to chemical fumigants, soil amendments provide a healthier soil, as chemical fumigants "tend to reduce microbial populations along with the pathogens. In such instances, the pathogens often rebound in numbers to cause a higher level of disease than seen before."

3.5 Summary of Amendment Benefits

As shown above, soil amendments provide multiple benefits in agricultural applications, and in many cases, the benefits are interrelated. For example, the lower soil bulk density associated with soil amendment use improves infiltration of water into the soil, which can reduce erosion. Reduced erosion helps keep topsoil and organics in place, improving soil productivity and, ultimately, crop yields.

The extent of the benefits provided depends upon many factors but, generally, literature shows that soil amendments:

- improve soil physical properties compared to conventionally fertilized and unfertilized plots: There is growing evidence that, although mineral fertilizers effectively provide nutrients for crop production, soil quality is poorer where they are used (compared to, for example, animal manures) (Stewart, Robinson, and Barker, 2000). Soil amendments, on the other hand, increase soil organic content, improve soil structure and water retention, and increase CEC. Moreover, soil tilth and quality improvements can be long-lasting.
- increase total soil microbial populations;
- provide primary and secondary nutrients, as well as micronutrients: The micronutrients contained in amendments can be especially beneficial, as farmers can reap the advantages of micronutrient addition without applying supplemental micronutrient mixes to their fields.
- can result in yields as high or higher than conventionally fertilized fields: Higher yields reported have been attributed to a combination of the nutrients supplied by the amendments and/or, improved soil conditions (including in some places, an improved pH);
- can reduce erosion; and
- can suppress some plant diseases (shown for composted materials and some non-composted manures).

The key to reaping these benefits is proper application. Any material improperly applied or applied in excess of recommended rates can be detrimental to the environment.

3.6 Mineral Fertilizer Benefits

Literature on the benefits of fertilizers to crops and soils is too extensive to be comprehensively addressed here. Therefore, this chapter briefly summarizes these benefits, with emphasis on contemporary issues.

Fertilizers supplement essential plant nutrients that soils cannot supply at levels required to support high-yielding crops. Fertilizers foster the ability of our most productive soils to achieve their maximum biomass-producing capacities. This has resulted in an increase in U.S. crop yield over the last 50 years, while cropland acreage has been declining from urban sprawl and other land use changes. Studies have shown that fertilization has compensated, in most cases, for wind and water soil erosion, and that increased biomass production has reduced erosion rates. On our most productive soils, crop productivity is often limited by drainage, and in the future, it is likely to be more limited by water supply than by soil fertility.

Some of the major benefits of mineral fertilizers versus other sources of nutrients are flexibility, specificity, and concentration. This is particularly true today with the development of precision farming. Fertilizers can be formulated to satisfy the exact nutrient needs of crops and livestock. They also have become increasingly concentrated (i.e., there is very little unused material or filler). With precision farming, fertilizer rates can be adjusted to reflect differences in soil test fertility levels, drainage patterns, soil physical properties, and historical crop yields. Fertilizers also have the flexibility to supplement nutrients that animal manures and biosolids would otherwise supply.

Much proclamation has been made in recent years about the superiority of organic farming over farming with mineral fertilizers and pesticides. Aside from the issue of pesticide use, it has been shown that organic farming produces no measurable differences from farming with fertilizers in terms of crop yield and quality, and livestock health and production, as long as the total nutritional needs of the crop or animal are met and organic matter is managed appropriately.

One of the justifiable criticisms of mineral fertilizer use, besides the tendency for over-fertilization, is its tendency to deplete the soil of non-applied nutrients and reduce the organic matter content of soils. Where crop residues are returned to the soil, no-till farming is practiced, or organic matter in the form of manure or biosolids is applied, however, comparable soil productivities are maintained.

CHAPTER 4.0

ENVIRONMENTAL AND HEALTH CONCERNS

Environmental and health concerns have been raised regarding each of the products we have discussed. Some of the concerns are specific to single products, but others — especially nutrient impacts — apply to both soil amendments and fertilizers that are improperly applied.

In this chapter, we explore concerns related to nutrients and metals for all products, and for soil amendments, pathogens and organic matter as well. Potential impacts to water quality, soil, and human and ecosystem health are addressed.

Increasingly, risk assessments have been used in the United States for quantifying concerns regarding materials applied to the land. For biosolids, and most recently, for fertilizers in California, they have provided a basis for regulation. To date, risk assessments have been performed for biosolids, MSW compost, and mineral fertilizers. A brief discussion of these assessments and their outcomes is presented below.

Where risk assessments have yet to be performed (or are called into question), it is critical to remember the difference between concerns (i.e., perceived risks) and actual risks. For example, some scientists are concerned about metal impacts from biosolids applications (Bouldin, 1997; Richards et al., 1998), yet even they admit that some application scenarios supporting their concerns may not be typical. Discussing a site that they used for research into biosolids impacts on metals in groundwater, Richards et al. (1998) noted that "because of a number of factors, including prior lead arsenate pesticide use, a single, extremely heavy sludge application, and loading rates of several metals in excess of USEPA cumulative loading limits, this site cannot be considered representative of application sites where sludge that meets regulatory limits is applied at agronomic rates."

Similarly, other researchers assessing metals transport from biosolids-applied metals (Camoreco et al., 1996) note that their findings "may not be directly applicable" to biosolids applications because of the methods they used. Please note that the concerns that these researchers raise bear consideration and, especially, their research demonstrates the danger of improper applications. But do those concerns reflect an actual risk to farmers applying biosolids in accordance with regulations? And, if there is a risk of any kind, how does that compare to the benefits that can be derived from using biosolids or to risks from other land applied materials?

The information that this chapter presents is intended to identify concerns, present risks as quantified by others and, where possible, put some concerns in perspective. Most importantly, it is intended to help answer, to the extent possible, questions such as the ones presented above.

4.1 Risk Assessments

Risk assessment has become the tool of choice in the United States for determining the environmental impact of soil amendment products. The principal approach, which was developed by USEPA for biosolids risk assessment, involved a multi-step process (USEPA, 1995). Initially, contaminants to be assessed were identified through a screening process that eliminated candidate elements and compounds because of their low concentrations in biosolids and/or their low potential hazard. Environmental profiles were then developed for the remaining contaminants. The profile database included fate and transport, as well as plant and animal toxicity data.

The environmental profile database was then applied to a pathway analysis procedure developed by USEPA. Pathways (potential avenues of exposure) assessed for the biosolid risk assessment are presented in Table 4-1. The outcome of the pathway analysis was to calculate allowable soil contaminant loadings for each pathway. The allowable loading for the limiting pathway became the regulatory limit in the 40 CFR Part 503 biosolids rule. The biosolids risk assessment is well described in USEPA's *A Guide to the Biosolids Risk Assessments for the USEPA Part 503 Rule* (USEPA, 1995). Subsequent to the publication of the 503 rule, the same approach was used to conduct risk assessments for municipal solid waste compost (Logan et al., 1999) and fertilizers (The Weinberg Group Inc., 2000). These three risk assessments are compared below.

4.1.1 Biosolids

The biosolids risk assessment evaluated more than 200 contaminants from the 412 that were surveyed in the 1988 National Sewage Sludge Survey (USEPA, 1990a). An Intra-Agency Task Force narrowed that list to 50 contaminants to be further investigated. Ultimately, risk-based pollutant loading limits were developed for 10 metals and 12 organic compounds, mostly pesticides (pollutants for which limits were not developed were deferred for later investigations). All of the organic compounds were eliminated from the final rule (promulgated in 1993) because the risk levels were low compared to typical biosolids concentrations. Chromium was later rescinded from the 503 rule, the selenium concentration limit was made equal to the maximum allowable, or ceiling concentration, and the molybdenum concentration limit was dropped pending a reassessment of the limiting pathway (molybdenum-induced copper deficiency in livestock). Of the remaining nine metals, the limiting pathway for five metals (arsenic, cadmium, lead, mercury, and selenium) was direct ingestion; three metals (cadmium, nickel, and zinc) were limited by plant phytotoxicity; and molybdenum was limited by livestock eating forage or feed.

Recently, USEPA revisited pollutants that had been initially identified as candidates for regulation but had not been included in the original 503 rule. For this "Round 2" assessment, USEPA considered 31 pollutants and, based on a risk assessment of 10 exposure pathways, selected one family of compounds —dioxins — for regulation (the family includes dioxins/ benzofurans, and coplanar PCBs). In December 1999, USEPA issued proposed dioxin limits for public comment. As noted in Chapter 2, the vast majority of biosolids easily meet these limits.

Table 4-1
Summary of Exposure Pathways Used in the Risk Assessment for Biosolids Land Application

Pathway	Description of Highly Exposed Individual
1. Biosolids>Soil>Plant>Human	Human (except home gardener) lifetime ingestion of plants grown in biosolids-amended soil
2. Biosolids>Soil>Plant>Human	Human (home gardener) lifetime ingestion of plants grown in biosolids-amended soil
3. Biosolids>Human	Human (child) ingesting biosolids
4. Biosolids>Soil>Plant>Animal>Human	Human lifetime ingestion of animal products (animals raised on forage grown on biosolids-amended soil)
5. Biosolids>Soil>Animal>Human	Human lifetime ingestion of animal products (animals ingest biosolids directly)
6. Biosolids>Soil>Plant>Animal	Animal lifetime ingestion of plants grown on biosolids-amended soil
7. Biosolids>Soil>Animal	Animal lifetime ingestion of biosolids
8. Biosolids>Soil>Plant	Plant toxicity due to taking up biosolids pollutants when grown on biosolids-amended soils
9. Biosolids>Soil>Soil>Organism	Soil organisms ingesting biosolids/soil mixture
10. Biosolids>Soil>Soil>Organism>Soil> Organism>Predator	Predator of soil organisms that have been exposed to biosolids-amended soils
11. Biosolids>Soil>Airborne Dust>Human	Adult human lifetime inhalation of particles (dust) (e.g. tractor driver tilling a field)
12. Biosolids>Soil>Surface Water>Human	Human lifetime drinking surface water and ingeseting fish containing pollutants in biosolids
13. Biosolids>Soil>Air>Human	Human lifetime inhalation of pollutants in biosolids that volatilized to air
14. Biosolids>Soil>Ground Water>Human	Human lifetime drinking well water containing pollutants from biosolids that leached from soil to ground water

Source: USEPA, 1995

Although the lack of demonstrated health impacts resulting from biosolids use appears to support the findings of the USEPA risk assessment, a recently released study suggests that the risk assessment be revisited (NRC, 2002). Specifically, the study notes that risk assessment methods have evolved in the decades since the 503 assessment was performed and recommends that a new risk assessment, incorporating new approaches and additional data be performed to "update and strengthen the scientific basis and credibility" of the 503 regulations.

4.1.2 Municipal Solid Waste Compost

Logan et al. (1999) assessed potential environmental risks from the land application of municipal solid waste (MSW) compost. The assessment used the pathway analysis from the biosolids risk assessment and published and unpublished data on contaminants in MSW compost. Plant uptake of metals and selected organics was calculated from the results of field and laboratory studies at the University of Washington, North Carolina State University, and The Ohio State University. Plant uptake from MSW compost relative to that of biosolids was used as a surrogate for contaminant transfers in the pathway analysis where no specific data was available for MSW compost.

The analysis showed that only lead occurred in MSW composts at concentrations as high or higher than biosolids; other metals were found in much lower concentrations in MSW compost than in biosolids. The metal boron is not regulated in the 503 biosolids rule, but is found in some MSW composts at concentrations high enough to be phytotoxic (Logan et al., 1999). Risks associated with MSW compost that were not addressed in the biosolids assessment are the presence of inerts (metal, plastic, etc.) and the potential for crop damage associated with N immobilization when composts are inadequately cured. The U.S. Composting Council is currently developing product standards that deal with some of these issues.

4.1.3 Mineral Fertilizers

Concerns about metal contamination of commercial fertilizers were heightened several years ago by a series of newspaper articles in the *Seattle Times* about the use of industrial byproducts in fertilizers. Subsequent surveys of commonly used fertilizers and fertilizer ingredients showed that, while these materials are for the most part low in trace elements, a few have high concentrations of one or more of them (see Chapter 2). These concerns have led to a call for national standards for metals in fertilizers, and a number of efforts have been launched to set such standards. In response to these activities, The Fertilizer Institute (TFI) contracted with The Weinberg Group of Washington, D.C. to prepare a comprehensive risk assessment of metals in commercial mineral fertilizers.

The assessment focused on phosphate fertilizers and micronutrient fertilizers, 11 metals (arsenic, cadmium, chromium, cobalt, lead, mercury, molybdenum, nickel, selenium, vanadium, and zinc) and the radionuclide Ra 226 (a contaminant in some phosphate rocks). Single and multi-crop farming scenarios were assessed, exposed populations were assumed to be farm families, and pathways of exposure included crop consumption, unintentional ingestion of fertilized soil, and dermal contact with fertilized soil. Risk was determined by comparing risk-based concentrations (RBCs) with maximum concentrations found in fertilizers. The fertilizer database currently includes 925 individual fertilizer samples and approximately 140 micronutrient fertilizer samples. The assessment has shown that none of the phosphate fertilizers have metal concentrations that exceed the RBCs, although several micronutrient fertilizers had concentrations of arsenic and lead that exceeded the RBCs (The Weinberg Group, 2000).

In 1999, USEPA also issued draft findings from its own risk assessment for mineral fertilizers, *Estimating Risks from Contaminants Contained in Agricultural Fertilizers* (USEPA 1999c). Exposure pathways examined include: direct ingestion of fertilizers (during their application); ingestion of fertilized soil; inhalation of particles and vapors during and after fertilizer use; ingestion of plant and animal products produced on fertilized soils; and the ingestion of fish caught in streams next to fertilized fields (USEPA, 1999c). USEPA concluded that, "based on the data available, hazardous constituents in fertilizers generally do not pose harm to human health or the environment." Based upon a few product samples, however, the study did note potential health risks from arsenic and dioxins in some liming agents and micronutrient fertilizers.

4.2 Nutrients

Although they are essential for healthy crops, nutrients (particularly nitrogen and phosphorus) can be harmful when applied excessively or imprudently.

Agricultural nutrient inputs from multiple sources have been implicated in several large-scale environmental problems across the country. For example, excess nutrients have been implicated in some outbreaks of *Pfiesteria piscicida* (a toxic aquatic organism) and associated fish kills in North Carolina estuaries, as well as in Virginia and Maryland tributaries to the Chesapeake Bay (USEPA, 1998). It should be noted, however, that some outbreaks in Maryland occurred in waters that did not have elevated nutrients (Kintzer, 2002). Human health impacts in researchers and people with heavy environmental exposure to *Pfiesteria* have also been observed (Grattan, et al., 1998). Mineral fertilizers and manures, which are the largest contributors of nitrate in the Mississippi River Basin, are suspected of enlarging a "hypoxic zone" (an area of depleted oxygen) in the Gulf of Mexico during the summer months where the Mississippi empties into the Gulf (Goolsbey and Battaglin, 1997).

The following chapter highlights studies of nitrogen- and phosphorus-related water quality issues for the multiple products investigated here.

4.2.1 Nitrogen

Nitrogen concerns generally focus on groundwater contamination, although surface water contamination can be a concern as well. Unlike phosphorus, which is strongly absorbed by most soil (Sharpley et al., 1998), nitrogen (as nitrate) moves through the soil with relative ease and therefore has a greater chance of entering groundwater.

Because groundwater can sometimes feed surface waters, nitrate in groundwater can adversely impact surface water quality, but it is primarily viewed as a health concern in drinking water supplies. It has been linked to methemoglobinemia ("blue-baby syndrome"), a condition that prevents the transport of oxygen in a baby's bloodstream (USDA, 1997). It has also been linked to abortions in livestock (CDC, 1996). To address human health concerns, USEPA established a maximum contaminant level of 10 mg/L for nitrate in drinking water, which is roughly four times lower than the tolerance level for cattle. Recent laboratory studies have found, however, that the 10 mg/L nitrate limit established for drinking water is not protective of other life forms, specifically amphibians (Marco, Quilchano, and Blaustein, 1999). Rouse, Bishop, and Struger (1999) report that nitrate concentrations as low as 2.5 mg/L have adversely impacted amphibians, and further report that elevated nitrates may impact the insects that amphibians feed on, as well as amphibian predators (fish).

The impact of agriculture on groundwater nitrate was illustrated by a random survey conducted by Monsanto (Sharpley et al., 1998). The survey found that water in 5% of the total wells sampled had nitrate concentrations greater than 10 mg/L, but the proportion of high nitrate measurements nearly doubled on farmstead wells. In an earlier, more comprehensive study, USEPA also estimated that water from rural wells was twice as likely to have high nitrate concentrations (e.g., greater than 10 mg/L) as community well water (USEPA, 1990b).

The susceptibility of nitrate to leach into groundwater is governed by many factors, including soil organic matter, clay content and soil permeability. But studies indicate that a considerable portion of the United States is at risk from nitrogen contamination of groundwater (USDA, 1997).

Multiple sources are responsible for nitrogen loads in our nation's rivers. As shown in Figure 4-1, manure and mineral fertilizers are primary contributors in some watersheds, although atmospheric deposition (from dust and the combustion of fossil fuels) and point sources (for example, wastewater discharges) contribute as well (Puckett, 1994). (Note that the study did not investigate biosolids, but their contribution to overall nutrients applied to the land is negligible when compared to fertilizers and manures).

Studies have shown that the overapplication of nitrogen can lead to nitrate contamination of groundwater. An Illinois program where biosolids were initially applied at a rate exceeding crop requirements found the nitrate in a groundwater well near the site to be generally above 70 mg/L and tile drainage nitrate concentrations of up 100 mg/L (Jacobs et al., 1993). Andres (1995) found that about one-third of the water supply wells in the area of intensive poultry production (and land application) in Sussex County, Delaware have nitrate concentrations exceeding 10 mg/L. Throughout the county, roughly 20% of the wells exhibited nitrate concentrations above the drinking water standard. Andres (1995) cites other studies showing that the high nitrate levels have also impacted surface water, resulting in the eutrophication of Delaware's Indian River Bay and many fresh water ponds in the area. He also predicts that because of the slow ground water flow and large area affected, significant nitrate contamination is likely to persist for decades, even if all nitrate inputs to groundwater cease immediately.

Studies focusing on biosolids and manures have shown that applying nutrients at agronomic rates dramatically reduces the potential for nitrate contamination. Groundwater monitoring for a 9-year biosolids application program in Virginia that used agronomic rates found an average of 6 mg/L nitrate in the well that exhibited the highest nitrate concentrations (Jacobs et al., 1993). A Wisconsin program found nitrate concentrations in downstream wells increasing over time, but a long-term study comparing trends at the site to trends at commercially fertilized and manured fields revealed that there were no significant differences in nitrate trends between the different nutrient sources (Jacobs et al., 1993). In a study of poultry litter applications on sub-surface nitrate, Daniel et al. (1995) found that nitrate in the soil increased with application rate, but that nitrate concentrations did not exceed 10 mg/L when the litter was applied at recommended rates.

Figure 4-1
Proportions of Nonpoint Sources in Nitrogen in Selected NAWQA Watersheds

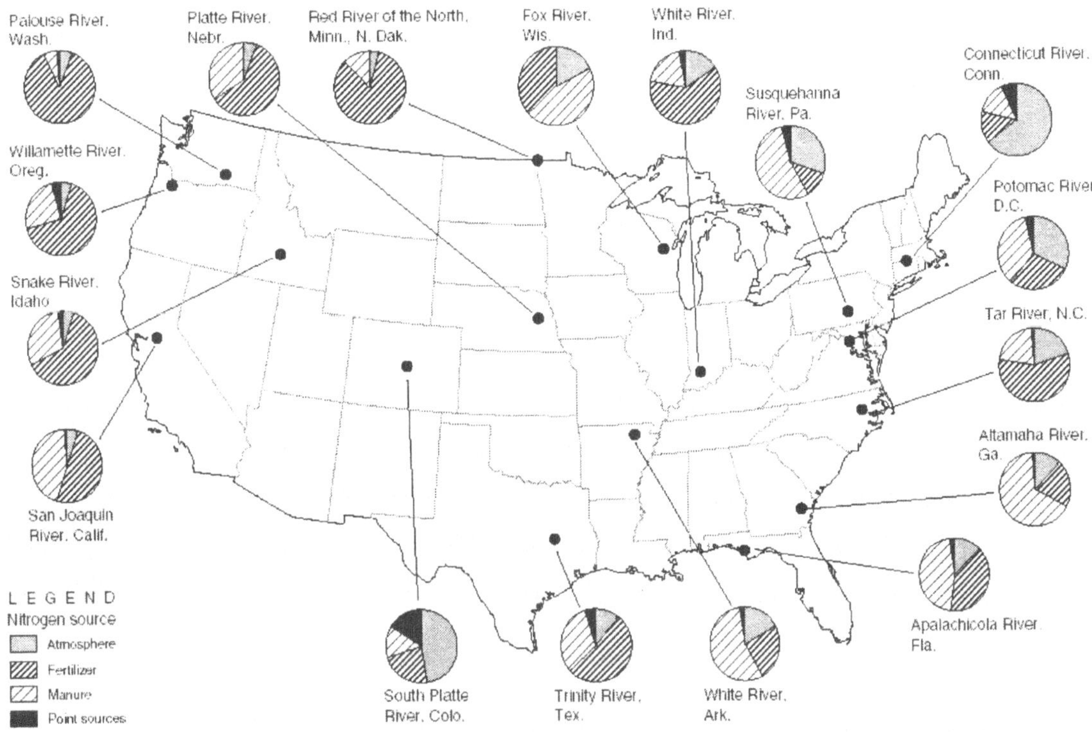

Source: Puckett, 1994

Other studies have compared nitrate impacts from amended and fertilized fields. Perhaps the most extensive study was conducted by USDA over a 20-year period at the Rosemount Experiment Station in Minnesota. Nitrate concentrations in shallow wells showed nitrates in excess of 10 mg/L for biosolids-amended fields and for fertilized fields, but nitrates were not elevated in deep groundwater (dense glacial till overlays the aquifer below the site) (Clapp et al., 1994). Nitrate levels decreased in recent years from biosolids-amended areas as application rates were decreased (but yields from these fields did not decrease). Comparing biosolids- and fertilizer-supplied nitrogen, monitoring shows that, in general, when biosolids N was oversupplied (1976–1981), excess nitrate was observed in groundwater. When biosolids-applied N was reduced, nitrate was generally greater in fertilizer wells (Figure 4-2).

One 3-year Swedish study found that leaching from poultry manure amended plots exceeded leaching from fertilized plots (Bergström and Kirchmann, 1999). In that study, however, manure nitrogen and inorganic nitrogen were applied at identical rates of 100 kg N/ha (89 lb/ac) for the first year of the study. In the following years, all plots (with the exception of control plots) received 100 kg N/ha/yr (89 lb/ac/yr) of inorganic N. This study, which did not credit nitrogen mineralization in the soil from a previous manure application, underscores the importance of considering the residual nitrogen from previous manure (or biosolids) applications when calculating N application needs.

Figure 4-2
Nitrate Concentrations in Shallow Groundwater

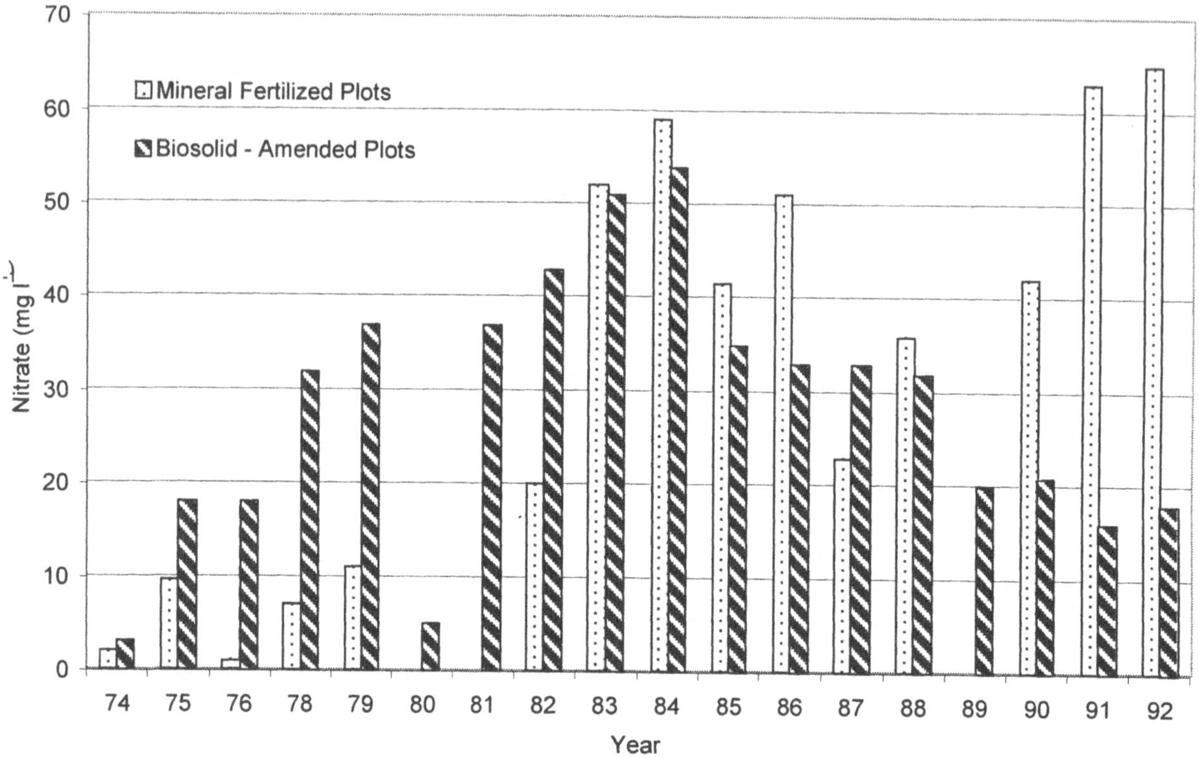

Source: Clapp et. al., 1994
Reprinted from Sewage Sludge: Land Utilization and the Environment with permission through the Copyright Clearance Center

The extent of nitrogen impacts on groundwater – especially for soil amendments - can also be significantly influenced by denitrification. One study (Lowrance et al., 1998) found that annual denitrification totals from liquid dairy manure applications ranged from 32% to 114% of the excess nitrogen available. Generally, it appears that denitrification from manure applications exceeds denitrification from fields where mineral fertilizer has been applied (Jokela, 1992; Loro, Bergström and Beauchamp, 1997). Where studies have found lower nitrate leaching from manure applications, as compared to mineral fertilizer applications, the differences have been attributed to the higher denitrification rates from manures (Jokela, 1992). Specifically, Jokela (1992) found that the application of manure resulted in similar or slightly lower soil profile nitrate than fertilizers when both products were applied at agronomically equivalent rates.

Studies assessing surface water impacts from amendment and fertilizer applications show mixed results. A comparison of runoff quality before and after digested biosolid applications at a site in Virginia showed slight increase in ammonia (and phosphorus) but a significant decrease in total nitrogen (Jacobs et al., 1993). At the Rosemount site, researchers found that surface water quality in a nearby stream was not degraded over the 18-year study period. Erosion control was carefully managed at the site, which benefited surface water quality (Clapp et al., 1994). In a series of studies, Daniel, et al. (1995) found that pasture-applied poultry litter, poultry manure, and swine manure can increase nitrate in runoff, but he did not find any noticeable improvements in runoff quality when inorganic fertilizers were used to supply nutrients.

4.2.2 Phosphorus

Unlike nitrate, phosphorus (P) is not directly toxic to humans and animals, but it is a leading cause of eutrophication (Sharpley et al, 1999), one of the primary reasons for impaired water quality in the United States (Sharpley et al., 1999), most especially in fresh water lakes and streams (nitrogen is usually the limiting nutrient for eutrophication in estuaries and other saltwater bodies).

Symptoms of eutrophication include the growth of algae and aquatic weeds (Figure 4-3), and oxygen shortages in the water caused by the death and decomposition of these plants. These phenomena restrict surface water use for fisheries, recreation, industry and drinking (Sharpley et al., 1999). Phosphorus (and nitrogen as well) has also been implicated in surface blooms of blue green algae (cyanobacteria), which can cause fish kills, drinking water taste and odor problems, and increase the production of trihalomethane (a carcinogen) during chlorination at water treatment plants. (Sharpley et al., 1998).

Figure 4-3
Eutrophication in Pond

Because P is not very mobile in soils, it can accumulate there. Sharpley et al. (1999) reports that a survey of agricultural soils by state soil test laboratories showed that a majority of these soils had high P levels relative to crop growth needs (Figure 4-4).

Unfortunately, many of the same areas that show high soil P are located close to sensitive water bodies, such as the Great Lakes, Lake Champlain, and the Everglades. Because runoff and soil erosion are the primary ways that P can enter surface waters, high soil P levels mean that these water bodies are very vulnerable to P overloads.

Figure 4-4
Percentage of Soil Samples with High Soil Test P

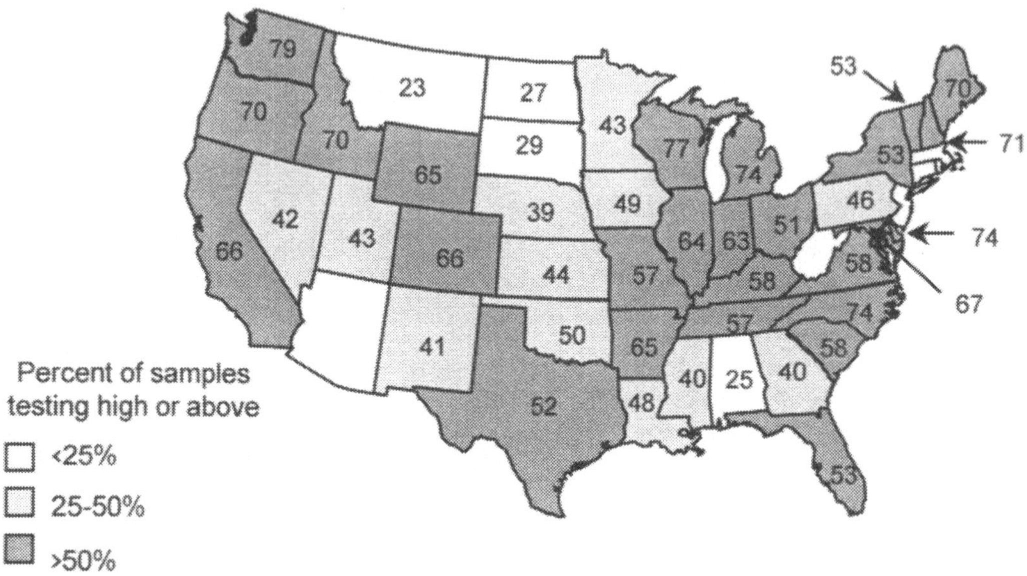

Source: Sharpley et. al., 1999

With respect to P accumulation, a specific concern with respect to soil amendments is their nitrogen-to-phosphorus ratio. Generally, biosolids and manures are applied at a rate that supplies sufficient nitrogen for crops. Because the N:P ratios in biosolids and manures are roughly 2:1 and 4:1, respectively, but crops need these nutrients only at a ratio of about 8:1, applying amendments at an agronomic rate for nitrogen can result in an oversupply of phosphorus.

The increased P levels from biosolids and manure application reflect this "unbalanced" nutrient supply. A survey of several long-term biosolids application programs (Jacobs, et al., 1993) found significant increases in available soil phosphorous due to biosolids applications that had been based on N requirements. Figure 4-5 shows how adding manure to meet N requirements dramatically exceeds crop P needs.

Excess soil phosphorus levels have been observed in both amended and fertilized soils, however. Based upon a multi-year study, Peterson et al. (1994) noted that biosolids and mineral fertilizer applications increased soil P to high levels (Table 4-2). Research by Sims et al. (1999) in the mid-Atlantic region supports P increases with biosolids applications but also found no indication that biosolids applications would increase the risk of P loss to waters relative to other P sources (fertilizers and manures).

Figure 4-5
Phosphorus Added versus Phosphorus Removed by Corn at Nitrogen-based Agronomic Rates

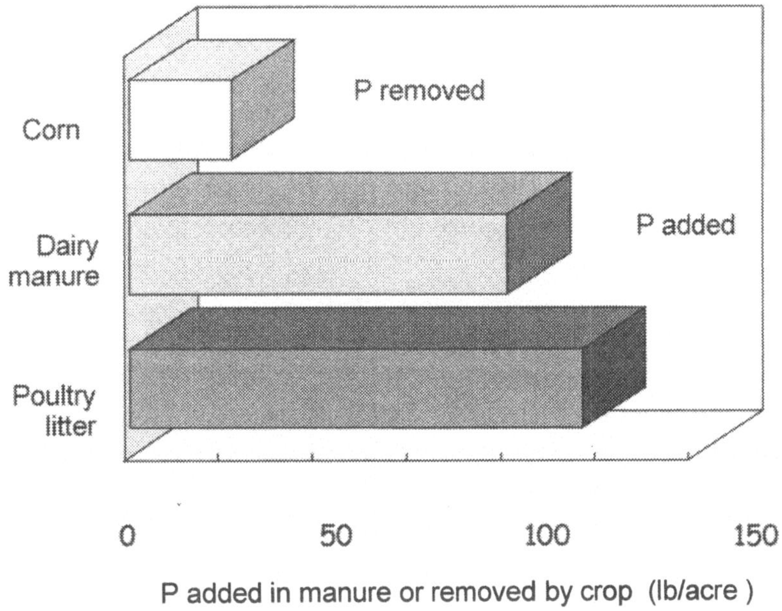

Source: Sharpley et. al., 1999

Table 4-2
Impact of Biosolids and Mineral Fertilizers on Soil Phosphorus in Elkhorn, WI

	Soil P (mg/kg)										
Treatment	**1981**	**1982**	**1983**	**1984**	**1985**	**1986**	**1987**	**1988**	**1989**	**1990**	**1991**
None	89	68	53	48	60	54	58	63	50	45	78
6.6.Mg/ha Biosolids	98	74	89	130	130	112	135	145	150	150	175
13.2 Mg/ha Biosolids	150	125	110	100	195	180	195	190	195	195	180
Mineral Fertilizer	--	--	--	78	99	100	130	145	120	135	200

Source: Peterson et al.,1994
Reprinted from Sewage Sludge: Land Utilization and the Environment with permission through the Copyright Clearance Center

Based upon limited studies, the runoff potential for amendments appears to be lower than for mineral fertilizers (although it should be remembered that amendments are often applied at a nitrogen-based rate, oversupplying phosphorus and increasing runoff potential). The lower relative availability of phosphorus in amendments (discussed in Chapter 2) may account for potential reduced impacts from this P source. Studies by Sims et al. (1999) suggest that the increases in soil aluminum and iron from biosolids applications might even reduce P runoff potential by increasing P sorption capacity. In a study of 193 soils, Sharpley and Sisak (1997) found that P availability in leachate from soils treated with inorganic phosphorus was consistently greater than soils treated with poultry litter. The lower P availability — and runoff potential — of poultry litter is attributed to the complexing of P with calcium, iron and aluminum.

Few data are available that directly compare P runoff from biosolids-amended, manure-amended, and chemically fertilized fields. Several studies, however, have shown that P in runoff from biosolids-amended fields is reduced compared to fertilized or manured fields. A 2-year British study compared phosphorus in runoff from fields amended with liquid-digested biosolids, dewatered biosolids, liquid cattle manure, and triplesuperphosphate (Withers, Clay, and Breeze, 2001). These materials were applied at maximum recommended P rates to a silty clay loam soil. The researchers found that fields receiving the triplesuperphosphate exhibited the highest P runoff, followed by cattle manure, liquid biosolids, and dewatered biosolids. In an earlier study, McLeod and Hegg (1984) also found that runoff from pastures amended with biosolids had lower P concentrations than runoff from manured or chemically fertilized fields. Busheé, Edwards and Moore (1998) found that a biosolids application (at one-half the agronomic rate for N) actually decreased runoff P compared to a control (no fertilization), while horse bedding (manure plus bluegrass) increased runoff P. For both amendments, however, runoff P was lower than the control runoff P when alum was added.

Some have taken advantage of P reactions with Ca, Al, Fe, and Mn, adding products that contain these materials (e.g., caliche, alum, fly ash) to reduce P availability in manures. One recent study (Dao, 1999) found that adding these amendments widened the effective N:P ratio by factors of 1.5 to 18. This approach may allow appliers to use manure to meet crop N needs without oversupplying P. Other studies have found that the addition of alum and aluminum phosphorus to manures can significantly reduce soluble phosphorus (Smith et al. 2001; Dao, et al., 2001).

4.3 Metals

Some metals, referred to as trace elements, are actually plant nutrients and are essential for healthy crops. Applied in excess of plant needs, however, there are concerns that these metals can harm water resources, crop yields or quality, and animal health. Other metals that are present in both amendments and fertilizer (for example, cadmium) are not strictly necessary for plant growth and there are concerns regarding the fate and impact of these metals on the environment as well. Research addressing these concerns in presented in this subchapter.

Like total metals content, research on the potential impact of soil amendment metals has focused more on biosolids than the other products included in this study and the depth of information presented here is weighted accordingly.

4.3.1 Leaching Potential

As noted in Chapter 2, metals availability — and correspondingly, leaching potential — depends upon many factors, one of which is the form of the metal. Based upon biosolids studies (studies of other materials are limited), other factors influencing leaching and availability include:

- ***pH:*** Metals are generally more soluble in acidic soils. At neutral and alkaline pH, most heavy metals will be in relatively insoluble forms, and leaching potential is diminished. This trend is generally supported for a variety of biosolids (Woodbury, 1992; Bennett, Gopalan and Senjupta (1989); additional research is needed to confirm that the same trend holds true for manure and mineral fertilizers as well.
- ***CEC:*** Cation exchange capacity is a measure of a soil's ability to exchange positively charged atoms or molecules with ions in solution. The addition of clay or organic matter can increase soil CEC (Epstein et al., 1976). Positively charged metal species are retained on the exchange complex to varying degrees depending on metal characteristics and other soil properties (e.g., pH).
- ***Organic matter:*** Organic matter added through the land application of biosolids provides additional sites for metals adsorption, and as noted above, increases soil CEC. Both of these factors can increase metals retention in soil and decrease leaching potential. The type of organic matter present may influence metal availability as well. Insoluble organic matter (as defined by humic acid content) inhibits the plant uptake of some metal cations; these cations bind strongly with the organic matter and are unavailable to plant roots (McBride, 1995) and are probably unavailable for leaching. On the other hand, soluble organics (as defined by fulvic acid content) form soluble metal-organic complexes (McBride, 1995), which are more available to plants and may be subject to leaching.
- ***Preferential flow*:** Some have suggested that preferential flow (through channels formed by worms or roots) could provide a means to transport metals through soil (Camoreco et al., 1996). The authors note, however, that their findings "may not be directly applicable" to biosolids-applied metals because they used metal salts, not biosolids, in their experiments.

With the exception of preferential flow, the above factors are often cited as the major influences on metals leaching potential in biosolids-amended soils, but biosolids-soil interactions are complex. Rather than examining the precise impacts of these factors, we looked to multiyear programs to assess the overall risk of metals leaching from biosolids (or other materials) into groundwater.

The many factors affecting metals release can hinder efforts to predict metals transport, but long-term field studies suggest that biosolids-applied metals do not migrate to groundwaters (Jacobs et al., 1993; Logan and Chaney, 1983; Smith, 1996; Sloan, Dowdy, and Dolan, 1998). In their study of long-term biosolids nationwide application programs, Jacobs et al., 1993, cited the following information. Fifteen years of groundwater monitoring at a land application site for

Chicago biosolids showed that biosolids applications did not impact groundwater quality, even though biosolids applied in the site's early years (mid-1970s) had relatively high metal concentrations compared to today. Groundwater monitoring associated with a Virginia site showed no increases in detectable metal concentrations above state standards. For cadmium and zinc, the frequency of groundwater standard exceedances measured in shallow wells actually decreased after biosolids were applied. In Albuquerque, NM, 5 years of groundwater monitoring showed an increase only in sodium and, due to the depth of groundwater at the site, that increase was not attributed to biosolids applications.

Rather than migrating to groundwater, many studies have shown that biosolids-applied metals accumulate in surface soils, generally to a depth of 2.5 feet (0.76 meters) or less (Jacobs et al., 1993; Dowdy et al, 1994;USEPA, 1977; Garcia and Mangaroo, 1994; Sloan, Dowdy, and Dolan, 1998). Baveye et al. (1999) reported on distribution of trace elements in a silt loam amended with anaerobically digested biosolids. Biosolids application significantly increased the total soil concentration of Cd, Cr, Cu, Ni, Pb, and Zn at 1.0-ft (30-cm) soil depths. Below the 30-cm depth, increased concentrations were observed to 2.5 ft, 1.5 ft, and 2.0 ft (75 cm, 45 cm, and 60 cm), respectively, for cadmium, zinc, and chromium. Barbarick, Ippolito, and Westfall, (1998) conducted a long term (11 years) study on the distribution of extractable metals (Cd, Cr, Cu, Mo, Ni, Pb, and Zn) in different depths of biosolids amended soils. The study site was amended with digested biosolids at 3.0 and 12 tons/ac (6.7 and 26.9 Mg/ha) application rates. Surface soils (>8 inches, or 20 cm) had the highest trace element concentrations. Berti and Jacobs (1998) investigated the distribution of trace metals (Cd, Cr, Cu, Pb, Ni, and Zn) in soils amended with biosolids from 1977 to 1986. This study showed that there was no metal movement below 1.0 ft (30 cm) of depth; however, these metals were found to be laterally distributed, probably due to tilling and other agricultural activities.

In some cases, researchers have not been able to "find" all biosolids-applied metals in soils (Jacobs et al., 1993; McBride, 1997). McBride (1997) suggests that these missing metals have may have leached over time through various means and that they may pose a threat to groundwater. He notes, however, that there has been no evidence of substantial leaching from the site on which his hypothesis is based (and where sludge was very heavily applied).

Because others had observed this phenomenon, Sloan, Dowdy, and Dolan (1998) undertook a comprehensive and controlled study to assess the fate of biosolids-applied metals. The researchers recovered essentially all of the cadmium, copper, nickel, lead and zinc applied from 16 years of biosolids applications in the top 2.0 ft (60 cm) of soil (See Box 4-1 for more detail on this study). Only 60% of the biosolids-supplied chromium was recovered, however, and the researchers attribute this to the chemical used to extract metals from the soil.

Box 4-1 Where Do the Metals Go?

To answer this question, researchers from the USDA and the University of Minnesota went back to the Rosemount Experiment Station, a study site where fertilization and cropping practices have been closely monitored and controlled for over 50 years (Sloan, Dowdy, and Dolan, 1998).

From 1977 to 1979, waste activated, dewatered biosolids were applied to the silt loam soil at Rosemount's I-9 site, which overlays sand and gravel to a depth of about 2 ft (0.6 m). Application rates were 27, 54, and 80 tons/ac (60, 120, and 180 Mg/ha). The biosolids were first surface applied and then immediately incorporated into the surface of the soil to a depth of 6 in. (0.15m). After that period, the field was maintained in agricultural production.

Cumulative biosolids and heavy metal loadings were recorded over the three-year application period to establish a baseline heavy metal loading to the soil. This monitoring was critical to later determine the fate of the biosolids applied metals.

Sixteen years after the biosolids applications, the soils were analyzed at depths of 0-0.5 ft., 0.5 ft-1 ft, 1 ft-1.5 ft (0-0.15, 0.15-0.30, and 0.30-0.45 meters) to determine the concentrations of metals in the soil compared to the total amount applied during the biosolids applications years before, as well as metal

Figure 4-6
Highest Biosolids Loading Rate

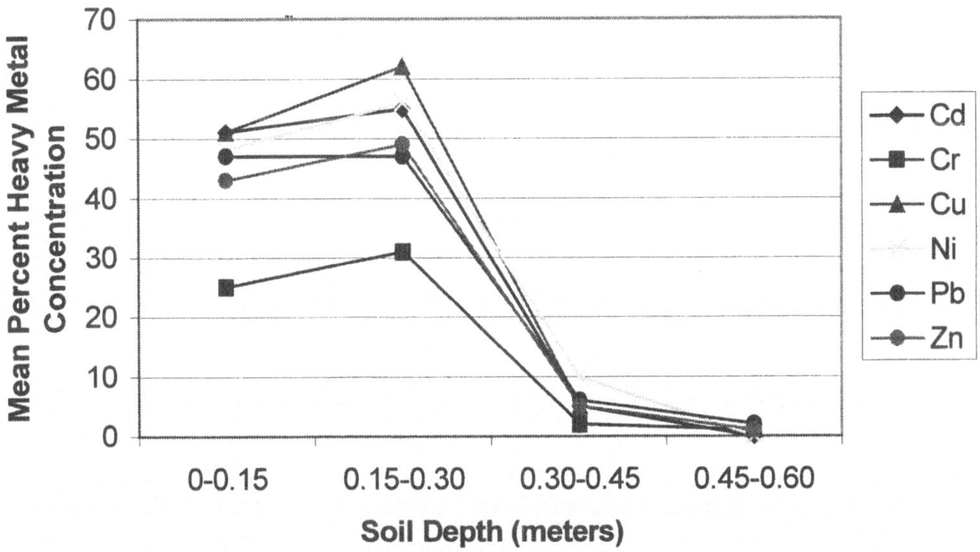

distributions in the soil. The results showed that greater than 90% of the cadmium, copper, lead, nickel, and zinc remained in the upper 1 ft. (0.30m) of the soil while 100% was present in the upper 1.5 ft. (0.45 m). Only 60% of the biosolid-supplied chromium was recovered, however, and the researchers attribute this to the chemical used to extract metals from the soil. The graph above illustrates the vertical distribution of recovered metals for the highest biosolid application rate. These results indicate that heavy metals migrate very slowly into the deeper soil and therefore pose little risk to groundwater contamination in this particular type of soil.

Studies comparing the leaching potential (and availability) of metals applied in amendments and fertilizers are limited. To date, only studies comparing metals movement in soil from biosolids and fertilizer applications have been identified. In a Colorado study, Barbarik, Ippolito, and Westfall (1998) found that biosolids-applied metals were generally found in greater concentrations in surface soils (compared to fertilizer-applied metals), but that this phenomenon was less pronounced deeper in the soil (zinc, which was present at quantities more than 100% greater in biosolids than in the fertilizer, was an exception to this finding). For some metals, in fact, soil metal concentrations at 8–24 inches (20–60 cm) depth from fertilizer applications were similar to metals from biosolids applications, despite the fact that the quantity of biosolids-supplied metals was greater. This phenomenon is illustrated in Table 4-3 for copper.

Table 4-3
Impact of Biosolids and Chemical Fertilizer Additions on Soil Copper in Colorado

		Soil Copper (mg/kg)			
Depth (cm)	Treatment	East B	West B	East A	West A
0-20	None	2.7	2.1	2.0	3.2
	Mineral Fertilizer(1)	2.4	2.3	1.9	3.6
	6.7.Mg/ha/application(2)	4.6	9.9	6.0	8.2
	26.8 Mg/ha/application(2)	9.4	8.3	11.9	13.9
20-60	None	1.6	1.8	2.0	2.4
	Mineral Fertilizer(1)	1.9	1.7	2.0	2.2
	6.7.Mg/ha/application(2)	1.8	1.8	2.5	2.2
	26.8 Mg/ha/application(2)	1.9	1.9	2.4	3.1
60-100	None	2.2	2.1	1.7	1.4
	Mineral Fertilizer(1)	2.0	2.1	1.8	1.5
	6.7.Mg/ha/application(2)	2.5	2.2	1.7	1.7
	26.8 Mg/ha/application(2)	2.4	2.2	2.0	2.6
100-150	None	1.9	1.7	1.4	1.2
	Mineral Fertilizer(1)	1.7	1.8	1.4	1.2
	6.7.Mg/ha/application(2)	2	1.8	1.2	1.3
	26.8 Mg/ha/application(2)	2.1	1.8	1.4	1.7

From: Barbarick, Ippolito, and Westfall (1998)

(1) 56 and 67 kg N/ha for sites A and B, respectively.

(2) 6.7 Mg/ha was recommended agronomic rate; 26.8 Mg/ha an excessive rate for study purposes.

In reviewing the values in the table, it should be noted that researchers estimate that almost 50 times more copper was applied in the biosolids than the fertilizer. Working with a heat-dried biosolids in Texas, researchers found that cadmium, lead and nickel concentrations in the soil (at 1.0–1.5 ft, or 30–45 cm depth) were greater with annual applications of commercial NPK fertilizer than they were with biosolids applications (Garcia and Mangaroo, 1994), although surface soil (0–6 inches, or 0–15 cm) concentrations were greater where biosolids had been applied. Like the Colorado study noted above, the Texas study also indicates that, with the exception of zinc, the amount of biosolids-applied metals decreases dramatically with depth.

Some studies indicate that metals from mineral fertilizers may also accumulate in the soil. The accumulation of cadmium from phosphate fertilizers has been a particular focus. Smith (1996) cites several studies that have found phosphate fertilizers to be the principal cause of cadmium accumulation in agricultural soils in the European community (EC). He specifically cites Van der Voet et al. (1994), who found that 50% of the total annual soil cadmium load to soils in EC countries was contributed by fertilizers, 3% by biosolids, 14% from atmospheric deposition and 18% from manure applications. Total cadmium contributions to the countries' soils are estimated to be 634 metric tons (575 tons) per year, with an average accumulation of 403 metric tons (366 tons) annually, or the equivalent of 3 g/ha/yr. In a California study, Mulla, Page, and Ganje (1980) also noted that most phosphate-applied cadmium is retained in the soil (71% in the top 6 in., or 15 cm).

Other studies indicate that manure, rather than inorganic fertilizers, may contribute greater quantities of cadmium to agricultural soils. (Whereas some metals, such as copper and zinc, are essential minerals for livestock and are deliberately added to livestock feeds, cadmium is usually introduced involuntarily via feed phosphates [Food and Agriculture Organization, 1999.]) In an Australian study of soils from 29 farms, Jindasa et al. (1997) found that soil cadmium almost doubled (to 14 mg/kg) from agricultural inputs, and concluded that heavy additions of poultry manure were responsible for most of the increases noted (although both manure and phosphate fertilizers were significant contributors). Reviewing data from the Rothamsted Experiment Station in England (perhaps the world's oldest agricultural study site, with more than 140 years' worth of data), Jones, Symon, and Johnston (1987) note that manure is likely a greater contributor to soil cadmium than inorganic fertilizers.

Studies have also shown that manure-derived zinc accumulates in surface soils (Jindasa et al., 1997; Chang, Sommerfeldt, and Entz, 1990; Christie and Beattie, 1989).

Although numerous studies indicate that biosolids-applied metals are retained in the soil, some researchers worry that these metals will ultimately be "released," becoming available for leaching and/or plant uptake. This belief is commonly known as the "time bomb" theory. The foundation of the theory, however, is that the organic matter in biosolids is largely responsible for the immobilization of metals in biosolids-amended fields. One hundred years of evidence refutes this belief, however (USEPA, 1995). Moreover, greater than 50% of the solids in biosolids are inorganic, and it is the inorganic biosolids fraction that exhibits some of the strongest metal binding characteristics (USEPA, 1995).

4.3.2 Plant Uptake

Per Logan and Chaney (1983), plant uptake of metals is important for three reasons:

1. Many metals are actually micronutrients and are necessary for healthy crops.
2. Excessive uptake could result in phytotoxic effects and reduced plant yield.
3. Plant uptake for some elements is a primary pathway for food-chain risk.

This chapter focuses on *phytotoxicity* (the retardation of plant growth) and potential human health impacts. Impacts on animal and aquatic species' health are discussed in subchapter 4.3.4.

Phytotoxicity can be difficult to assess because a number of variables - real or apparent can impact plant health, including: cold weather, salts, improper nutrient supply, high soil metals, pesticides, herbicides, and air pollution (USEPA, 1995). Even with proper management, yields can vary by 15% to 25%. Nonetheless, researchers have attempted to assess the impact of metal uptake on plants from biosolids-amended, mineral fertilized and, to a far lesser extent, manure-amended fields. Information collected to date is presented here.

To a large extent, research on phytoxicity and potential yield reductions has focused on biosolids, and, particularly, on the metals cadmium, zinc, copper, and nickel.

Of late, specific concerns have been expressed regarding the protectiveness of USEPA 503 rule for crops (McBride, 1995). Using zinc impacts on corn yields as an example and some of the same data used to develop the 503 rule, Bouldin (1997) concludes that once corn leaf zinc concentrations reached 100 ppm (at a cumulative zinc soil load of over 200 kg/ha, or 7% of the USEPA allowed maximum of 2,800 kg/ha), crop yields were reduced compared to non-biosolids amended plots. As noted previously, many factors can impact crop yield and whether or not any of these factors came into play is uncertain. Moreover, Bouldin notes that "loadings approaching the USEPA limits can only be attained by adding sludges very badly contaminated with metals and/or adding amounts of sludges far beyond that needed to supply the economic amounts" of nutrients required for crop growth. (With the current emphasis on proper management of biosolids, and intense public scrutiny of biosolids management in general, neither of these scenarios is likely.)

More important than extrapolations such as the above, perhaps, is the data obtained from over 600 cumulative years of biosolids applications, and limited studies of impacts from fertilizer and manure-supplied metals. These studies indicate that the uptake of metals from soil amendments and fertilizers has not adversely impacted crop yields. Evaluating 10 long-term biosolids application sites across the nation, researchers saw no phytotoxic effects from metals loadings, even at a site that received biosolids applications exceeding 503 Cumulative Pollutant Loading Rates for six metals (Jacobs et al., 1993). (The researchers noted that models used for 503 development would predict the observed outcome of no physical toxic effects.)

Other long-term studies evaluating digested biosolids (Clapp et al. 1994), alkaline-stabilized biosolids (Hormann et al., 1994), and heat-dried biosolids (Mangaroo and Garcia, 1994) found yields similar to chemically-fertilized plots although soil metal concentrations and plant uptake of some metals was greater in biosolids-amended fields.

Why? What factors are at work here that appear to protect plants from metals supplied by amendments and fertilizers?

A critical factor is soil pH. For some metals, soil pH impacts plant uptake (Chaney, 1994; van der Watt, Sumner, and Cabrera, 1994; Mortvedt, 1985). Chaney (1994) notes that strongly acidic soils increase the plant uptake of zinc, cadmium, nickel, manganese, and cobalt, and they also increase the potential for copper, zinc and nickel phytotoxicity. However, the agricultural use of acidic soils (pH less than 5.5) is unlikely, because such soils must usually be limed to a pH of at least 6.0 to prevent naturally occurring soil metals – such as aluminum and manganese – from hindering plant growth (USEPA, 1995).

For biosolids, several mechanisms protect both humans and animals from metals taken up by plants (USEPA, 1995). These mechanisms traditionally are referred to together as the "soil-plant barrier." First, some metals are so insoluble or strongly adsorbed in soils (chromium, for example) or plant roots (lead) that they are not transferred to the edible portions of plants (even if their concentrations are high in biosolids-amended soil). Iron oxides, aluminum, manganese, and phosphate can each bind metals tightly in the soil. Secondly, if high-metal biosolids were to be applied to soils with a pH at or below 5.5, copper, nickel and some other metals (along with natural soil metals) could result in unhealthy and small plants. These plants would be less desirable for consumption and could be present in lower quantities than healthier crops, decreasing their potential impact on human or animal consumers.

For biosolids, studies indicate that the multiple elements present in these materials can also have a protective effect on plant consumers (USEPA, 1995). For example, biosolids can supply cadmium for plant uptake, but they also provide zinc (at a ratio of about one part cadmium for each 100 parts zinc), iron and calcium. When an animal eats crops grown on biosolids-amended soils, they ingest these other metals in addition to cadmium, and these compounds actually work to decrease the amount of cadmium adsorbed in the intestine. They also decrease the adsorption of selenium and molybdenum. This phenomenon is addressed in detail in USEPA's "A Guide to the Biosolids Risk Assessments for the USEPA Part 503 Rule" (1995).

The beneficial balance and synergistic effects of elements on plant uptake have also been noted for mineral fertilizers and manures. Although zinc and cadmium can be present in high concentrations in some manures, phosphates fertilizers and zinc micronutrients, several studies indicate that (as with biosolids), soil phosphate limits the uptake of cadmium and zinc (Jindasa et al., 1997; Mulla, Page, and Ganje, 1980). Potentially supporting these studies, Mortvedt (1985) found that the cadmium uptake of several vegetables grown using triple superphosphates (containing 4 to 28 mg/kg cadmium) was no different than the uptake of plants grown on unfertilized fields. Indicating a beneficial relationship between cadmium and zinc (again, paralleling biosolids studies), research on wheat grain has shown that zinc micronutrient additions can actually decrease cadmium uptake from zinc deficient soils (Oliver et al., 1994).

The above studies indicate that high soil metals cannot be directly correlated to toxic conditions for plants, humans or animals. For example, studies have shown that foods grown in soils with very high cadmium levels (100 mg/kg–360 mg/kg from nearby mining and zinc smelting operations) did not impact the health of people eating those foods (ASTDR, 1994; Morgan and Simons, 1988). In these studies, researchers found that the cadmium:zinc ratio was 1:100 or more, once again demonstrating the synergistic and beneficial relationship between these metals.

This finding does not imply that heavy metal loadings should not be carefully managed. Christie and Beattie (1989) found that heavy annual applications of liquid swine manure at 2,400 gal/ac (22 m^3/ha) over a 16-year period increased copper concentrations in herbage to 10 mg/kg, a level that is nearly toxic to sheep. (The authors note, however, that lower application rates more typical of farm practices did not produce this effect.) Chaney (2001) also reports that heavy applications of swine manure to acidic coastal plain soils caused zinc toxicity to sensitive peanut crops in North Carolina. Once the pH of the soil, which had fallen, was restored to a level appropriate for peanuts, the adverse impacts were eliminated.

In Australia, excessive annual additions of poultry litter and traditional fertilization practices over many years increased cadmium concentrations to, in some vegetables, unacceptable levels of 0.05 mg/kg, the maximum pollutant concentration established for Australia vegetables (Jindasa et al., 1997). To reduce soil and plant cadmium, the researchers recommend the adoption of several management techniques, including: reducing litter and fertilizer cadmium concentrations, maintaining neutral to alkaline soil pH, ensuring an adequate zinc supply, salinity control, and the cultivation of crops that do not easily accumulate cadmium.

With respect to amendments other than biosolids and manures, it should be noted that only one metal — boron — has been identified as a potential phytotoxic concern (in MSW composts). In a study assessing metal impacts from MSW composts, Logan, et al. (1999) found that the 503 limits set for biosolids would be sufficiently conservative for the use of MSW compost, but that boron-induced phytotoxicity might be a concern. Woodbury (1992) also found a potential for boron toxicity but found that uptake of other metals from MSW compost-amended soils was limited.

In sum, it appears that concerns regarding plant uptake must consider not only how much metal is applied to the soil, but how available the metal is for uptake considering soil characteristics and the characteristics of the material applied. Additionally, plant type must be considered when weighing the risks of plant uptake, as metals tend to accumulate in plant leaves, and so are a greater concern for plants where the leaves are consumed (lettuce, for example, as opposed to corn grain).

4.3.3 Microbial Activity

Research on the impact of amendment or fertilizer-supplied metals on microbial activity are limited and appear to have focused on biosolids and composts.

Most studies involving biosolids have shown no adverse impacts to soil microbial activity (USEPA, 1995) from biosolids-supplied metals. For example, one study in England had indicated that soil metals may have caused a strain of *Rhizobium* to lose its nitrogen fixing abilities on one strain of white clover. (*Rhizobium* is an organism that fixes nitrogen; that is, it converts nitrogen in the air to organic nitrogen. It is important to both agriculture and the environment). USEPA reports that subsequent studies indicate that low pH may have been the cause for the observed problems, since the phenomenon was observed in control and biosolidS-amended plots. Another study, assessing the effect of biosolids compost on *Bradyrhizobium japonicum*, the microsymbiont of soybeans, found that the negative impacts on *B. japonicum* populations were caused by salinity, not by heavy metal content (Madariaga and Angle, 1992).

Some studies using biosolids and manures with very high metal contents have shown a reduction in microbial activity. Heckman, Angle, and Chaney (1987) found that the application of a biosolid containing 4,400 mg/dry kg of zinc and 2,200 mg/dry kg of copper did reduce nitrogen fixation. (40 CFR 503, Table 3 limits for zinc and copper are 2,800 mg/dry kg and 1,500 mg/dry kg, respectively.)

Another study by Moreno, Hernandez, and Garcia (1999) did find that the application of a high-cadmium compost to arid soil inhibited microbial activity, but that the benefits from the organic matter applied compensated for the inhibition.

For manures and biosolids and mineral fertilizers, researchers have noted a strong correlation between soil pH and N-fixing microorganisms (Mårtenssen and Witter, 1990; Ibekwe et al., 1995). Specifically, they note that low pH can be detrimental to these microbes' activities. Although Mårtenssen and Witter (1990) were also concerned about metal impacts from biosolids applications, they noted that "fertilization with organic fertilizers, for example, directly increases the carbon input into the soil, thereby increasing the size of the soil microbial biomass which can be supported. Long-term application of acidifying fertilizers, on the other hand, will impair the survival capability of many soil microorganisms as a result of a reduction in soil pH."

4.3.4 Animal and Aquatic Species' Health

Literature identified to date focused primarily on animal health. No specific studies (with the exception of *Pfiesteria*, a nutrient issue) have looked at metal impacts on aquatic species, and research on terrestrial impacts is limited as well. As has been the case for metals, most research has addressed biosolids.

Animals can be exposed to pollutants in biosolids through biosolids particles adhering to plants, biosolids on or in the soil, and pollutants taken up by plants that are then ingested (NRC, 1996).

Little comprehensive research has been performed, and most studies have focused on individual species impacts. Based on these studies, it appears that most trace elements do not accumulate in animal muscle tissue, although some may accumulate in liver and kidneys (NRC, 1996). Bray et al. (1985) found that liver and kidney cadmium and zinc were elevated for goats fed silage grown on biosolids-amended soils for 3 years, but muscle tissue was not affected. Neither was animal performance affected (e.g., milk production per kg feed consumed or rate of gain per kg feed consumed (lambs) (Dowdy et al., 1983). The lack of tissue accumulation was noted with cattle (Bertrand et al., 1981), swine (Hansen and Hinesly, 1979), and mice or guinea pigs fed Swiss chard (Chaney, 1978e; 1978b).

One study of biosolids compost and digested biosolids did find tissue accumulation for zinc. Rabbits from a mine area reclaimed with Philadelphia Mine Mix, a 1:1 mixture of biosolids compost and digested biosolids, were found to have elevated zinc levels in muscle tissue (Dressler et al., 1986). The study concluded that the levels of zinc were not sufficient to impact human health in the case of occasional rabbit consumption, but more study on the effect of elevated zinc on habitual rabbit predators is necessary.

In reviewing the results of this study, it is once again useful to consider perspective and actual risk versus concern, especially in the context of the agricultural applications that are the focus of this study. Specifically, biosolids loading rates used in reclamation efforts are often orders of magnitude greater than agronomic rates and, accordingly, caution must be used when comparing the potential impacts of these two very different biosolids uses.

Based upon a 3-year study, O'Conner and McDowell (1999) found that the transfer of metals from biosolids applications to forage was minimal for a variety of application rates (the minimum rate studied was the agronomic rate), and that recommended metal concentrations in feed were not exceeded. Molybdenum, a focus of the study, never exceeded 3 mg/kg in forage and the copper: molybdenum ratio remained within acceptable levels (2:1). Concerns regarding molybdenosis (molybdenum-induced copper deficiency) were a driving force behind the study, and the cattle in the study did exhibit symptoms of copper deficiency. The high sulfur content of the forage grown on biosolid-amended soils, however, was cited as the cause of the copper deficiency, and copper supplements (given in the third year of the study) ameliorated the problem. The researchers recommended mineral supplements with copper to meet nutrition needs in animals grazing on such forage.

Other researchers have also investigated the potential relationship between biosolids and molybdenosis. In field and greenhouse studies, McBride et al. (2000) did find low copper: molybdenum (Cu:Mo) ratios in forages grown on a heavily sludged site and in soil columns that had reflected cumulative loading rates of 96 ton/ac (215 metric t/ha) of dewatered biosolids and 296 ton/ac (663 metric t/ha) of alkaline stabilized biosolids over a several-year period. In a survey of local farms that had applied an alkaline-stabilized biosolid (reportedly at 9 ton/ac or metric 20 t/ha), the researchers also found elevated Mo concentrations (compared to unamended fields) and relatively low Cu concentrations in farm forages. Contrasted with the above study, a field study in Pennsylvania using a Class A alkaline stabilized biosolids found that, even when this product was applied at twice the recommended liming rate, the lowest Cu:Mo ratio in alfalfa was 2.5:1 (Stehouwer, 2001). This lowest ratio was recorded with the first of three cuttings; in the second and third cuts, Cu:Mo ratios were 3.3 and 8.1, respectively.

As opposed to studying the effect of plant metals in forage or silage, some have looked at the impacts of direct biosolids ingestion. At varying proportions of biosolids in the total diet and, for most animals, no adverse effects were noted (NRC, 1996). An exception was baby chicks fed a diet composed of 20% biosolids.

Researchers at Oak Ridge Laboratories performed a risk assessment of forest and impacts from biosolids applications and found that a risk to wildlife populations from a one-time application of 18 tons/ac (40 Mg/ha) of biosolids was unlikely (Efroymson et al., 1998). The researchers did note concerns regarding potential zinc and copper impacts on white-tailed deer, but also qualified that concern by noting that data was limited for this conclusion. Other important qualifications must be added to this study when reviewing its findings. First, the document does not separately assess impacts from nutrients and potentially toxic compounds. Secondly, the modeled application rate far exceeds the recommended agronomic application rate and therefore does not represent typical application rates; and, critically, the assessment is based upon models only, without confirmation or verification of the findings through field studies.

Perhaps the most important conclusion of the Oak Ridge study is that additional research of ecosystem impacts is required. The study focused on biosolids, but many of the concerns raised could be applied to other materials as well, and these other materials (manures, fertilizers) should be included in subsequent studies.

4.4 Pathogens

As shown in Chapter 2, both biosolids and manures contain pathogens. The key difference between these materials is that biosolids are, by law, processed to significantly reduce pathogen content (to nearly undetectable levels for Class A products) whereas manures are held to no such standard.

The relative contribution of pathogens from biosolids and manures is an important consideration in determining environmental impact. In England, where pathogen reduction requirements for biosolids are not as strict as the United States, manures contribute considerably more *Salmonella* to the environment than biosolids do. The same is true in the United States, however, the sheer volume of manures applied, coupled with the propensity for herd or flock infection, are cited as the basis for higher pathogen contribution of manures (WaterUK, 2000).

Other pathogen sources must be considered as well. Wild animals and birds contribute to ambient pathogen levels through their manures and even infected carcasses (WaterUK, 2000). For example, one study found that, near a well-operated livestock operation, manures from about one-third of the wild animals trapped contained *Cryptosporidium* oocysts (Drury and Lloyd, 1997).

To date, no cases of pathogen-related health impacts from biosolids have been documented, but manures have been implicated in several pathogen-related outbreaks in North America. Table 4-4 summarizes recent outbreaks that have been attributed to manure applications.

The potential for amendment pathogens to survive outside their host animal and to contaminate water supplies or crops depends upon many factors, including:

- Presence, quantity, and viability of the organism;
- Method of application (i.e., surface application or incorporation);
- Amendment solids content (i.e., whether the materials are applied as a liquid or solid);
- Soil conditions such as permeability, infiltration, soil moisture, texture, organic matter, and pH;
- Climatic exposure, especially temperature and sunlight; and
- Soil biological conditions (i.e., competitive organisms).

Researchers have found that many pathogens do not survive for extended periods (Akin et al., 1983; Rudolfs et al., 1950; Golueke, 1983; Lance, 1977; Sorber and Moore, 1986). Data supporting their findings are summarized in Table 4-5.

With respect to groundwater contamination, however, several researchers have noted the importance of soil structure (in some cases over pathogen survival periods) to pathogen transport (Stoddard et al., 1998; Goss et al., 2002; Peterson, Davis, and Orndoff, 2000).

Table 4-4
Examples of Recent Pathogen-Related Illness Outbreaks

Location	Year	Contaminant	Impact	Suspected Source	Reference
Walkerton, Ontario	2000	*E. coli*	7 deaths, 2,000 illnesses	runoff from manure applications entering water supply after torrential rains	O'Connor, D., 2002
Connecticut	1996	*E. coli*	Illness	cow manure used in organic lettuce growing	Nelson, 1997. Cited in EPA, 1998
Montana	1995	*E. coli*	Illness	cow manure used in organic lettuce growing	Nelson, 1997. Cited in EPA, 1998
Illinois	1995	*E. coli*	Illness	cow manure used in organic lettuce growing	Nelson, 1997. Cited in EPA, 1998
Maine	1994	*Cryptosporidium*	Illness (several hundred)	cider made from apples gathered from a cow pasture	Millard, et al. 1994. Cited in EPA, 1998
Milwaukee, WI	1993	*Cryptosporidium*	100 deaths, 403,000 illnesses	cow manure applications, slaughterhouse wastewater, wastewater treatment plant effluent	EPA,1998
Maine	1993	*E. coli*	Illness	manure applications to vegetable garden	Cieslak et al., 1993. Cited in EPA, 1998

Table 4-5
Pathogen Survival in Soils

Organism	Media	Survival (days)	Reference
Salmonella typhimurium	Soil	<28 - 70	Golueke, 1983
Salmonella typhi	Soil	30 - 120	Golueke, 1983
Fecal streptococci	Soil	23 - 67	Rudolfs et al., 1950
Tubercle bacilli	Soil	>180	Rudolfs et al., 1950
Poliovirus	Sand	<77 - 91	Golueke, 1983
Poliovirus 1	Loamy find sand	84	Golueke, 1983
Coxsackievirus B3	Clay	<161	Golueke, 1983
Enteroviruses	Sandy soil	90 - 170	Golueke, 1983
Enteroviruses	Loam	90 - 150	Golueke, 1983
Entamoeba histolytica	Loam and sand	8 - 10	Golueke, 1983
Ascaris ova	Sandy soil	<90	Golueke, 1983
Hookworm larvae	Sand	<120	Golueke, 1983
Hookworm larvae	Soil	42 - <180	Golueke, 1983
Giardia	Soil	49	Olson et al., 1999
Cryptosporidium	Soil	>84	Olson et al., 1999

Supporting empirical evidence (from manure-related illness outbreaks) that the risk of contamination from manure applications is greater than for biosolids, several long-term biosolids studies assessing indicator organisms have shown no impacts on groundwater or soils. Two long-term biosolids application sites (Madison, WI and Raleigh, NC) have monitored for bacteria in wells surrounding sites and detected no bacteria (Jacobs et al., 1993). Wells at the Madison site are tested for total coliforms while the Raleigh wells are tested for fecal coliforms. The Denver Metropolitan Water Reclamation District tests soil at its application site for fecal coliforms, also with negative results (Jacobs et al., 1993).

Although limited in number, studies of manure applications have demonstrated that bacteria (and parasites) can be transported to groundwaters and surface waters. A study by CDC (Campagnolo and Rubin, 1998) evaluated the groundwater and surface water potential contamination from large-scale swine operations in Iowa. *E. coli* and *Yersinia* were found in agricultural drainage ditches and wells, monitoring wells, tile outlets and a river. Levels of *E. coli* ranged from undetectable to 3,700 MPN/100 mL. *Enterococcus* sp. levels ranged from undetectable to 13,000 MPN/mL. *Cryptosporidium* was also detected in a river near the operation. In a study of poultry manure litter applications, Giddens and Barnett (1980) found that runoff from the application of poultry manure to fallow and grassland had fecal coliform levels as high as 3.8×10^6/ml.

In a Kentucky study, McMurry, Coyne, and Perfect (1998) found that the fecal coliform concentration in drainage water from fields amended with poultry manure consistently exceeded 200,000 fecal coliforms per 100 mL, and was as great as 30 million fecal coliforms per 100 mL in leachate. Looking at the impact of application method on bacteria transport, a Canadian study (Wall et al., 1997) found that regardless of application method, tile water from the fields amended with liquid swine manure was impaired within 7 to 30 minutes after application, exceeding water quality guidelines for bacteria, ammonia, and phosphorus for several hours. The day after manure applications, the fields were irrigated and tile water, once again, reflected increased nutrients and bacteria. The impacts appear to be short-lived, however, as the nutrients and bacteria fell below detectable levels within a few days.

Another Canadian study found an interesting relationship between manure handling methods and *Crytposporidium* contents in manures. In a study of fresh and stored manures on 60 farms, Fleming et al. (1997) found this parasite in 26.5% of all liquid swine manure samples but in only 8.1% of the solid dairy manure samples and 7.3% of the liquid dairy manure samples. Citing a link between oocyst findings in young animals and manure storage contamination, the researchers note that the mixing of all manures on swine farms may partly explain the greater number of positive tests at these facilities (at dairy farms, calf manures are handled separately, and the authors note that if these young animals are the primary source of oocysts, then keeping their manures separate would prevent contamination of the main storage tank).

4.5 Organic Matter

The amendment-supplied organic matter that is so valuable to soil can be detrimental to water quality. Microorganisms breakdown organic matter entering streams or lakes, a process which uses dissolved oxygen in the water and reduces the oxygen available for aquatic life (Minnesota Environmental Quality Board [MEQB], 1999; Ritter, 2001). Busheé, Edwards, and Moore (1998) found that horse bedding (manure plus bluegrass) and biosolids both increased the organic matter, as measured by chemical oxygen demand, (COD), in runoff from fields of fescue, although the increases were not high. For example, the COD in runoff from biosolids-amended fields was 36.2 mg/L compared to a control of 23 mg/L, while the horse bedding increased COD to 56.6 mg/L.

4.6 Other Chemicals

As Chapter 2 notes, each of the products studied here contains chemicals that are not needed for plant growth or, when present in excessive quantities, can be harmful. To our knowledge, no studies have demonstrated adverse impacts from these chemicals (dioxin in biosolids and mineral fertilizers, perchlorate in mineral fertilizers, organic compounds in biosolids).

However, chemical impacts from other sources have been noted to be potential concerns: salts from biosolid and manure applications and antibiotics from manures, clopyralid and picloram (herbicides) in yard waste compost (Bezdicek et al., 2001) and endocrine disruptors in biosolids and manures.

4.6.1 Salts

Salts in land-applied materials can accumulate in the soil, leading to phytoxicity (for some plants), and deteriorating soil quality by reducing permeability and contributing to poor tilth (USEPA, 1998). Additionally, salt runoff to fresh water systems can disrupt ecosystem balance and leached salts can impact the drinkability of water supplies (USEPA, 1998).

Salinity impacts have been noted with some biosolids, manure, and MSW compost applications.

A long-term study in Albuquerque, NM, found that, at high application rates, salts accumulated at their initial land application site, inhibiting the growth of native grasses, and allowing salt tolerant, early successional plants to take hold (Jacobs et al., 1993). Soluble salts at the site where biosolids were applied at an estimated 12–14 dt/ac (27–31 dry Mg/ha) ranged from 600 mg/kg–10,000 mg/kg at surface, 900 mg/kg–14,000 mg/kg at 1–2 ft (30–60 cm) depth. As noted in the study, "the problem in arid climate is salt buildup and changing flora of the site, rather than off-site transport of contaminants."

The above study supports NRC findings (based upon WWTP effluent leaching) that salts will accumulate in soils unless they are removed by plants or leaching (NRC, 1996). In temperate or humid regions, moisture will generally prevent salt buildup. In arid regions, however, salts will buildup unless enough water is provided to promote leaching.

Other studies have found some negative impact from biosolids-related salt accumulations. A study of the effect of biosolids compost on *Bradyrhizobium japonicum*, the microsymbiont of soybeans, found that the negative impacts on *B. japonicum* populations were caused by salinity, not by heavy metal content (Madariaga and Angle, 1992) as supposed. In a separate study, Abd El-Hay and Angle (1994) demonstrated (with anaerobically digested sludge) that soluble salts in biosolids-amended soils might restrict soybean growth at high sludge application rates (90–360 tons/ac or 200–800 Mg/ha, far above recommended agronomic rates).

These impacts can be reduced, however, if rainfall or irrigation washes salts from the root zone before planting.

With respect to MSW composts, researchers have found that this product inhibited plant growth in some trials. Dickerson (1996) reported losses of pepper plants due to high salinity in compost, and Gadjos (1997) found that germination of certain seeds was inhibited by high salinity in compost.

No detailed studies of manure-induced salinity were noted, but CAST (1996) reports that high application rates of cattle and poultry manures have adversely impacted soil permeability and water holding capacity in a fine textured soil.

4.6.2 Antibiotics

As noted in Chapter 2, antibiotics have been found in manures (and may be present in biosolids as well). The CDC study that detected antibiotics in manures (see Chapter 2) also detected antibiotics in at least one surface water sample, taken from a stream near a poultry house (Karpati and Rubin, 1998).

The implications of this finding are unclear, but it raises concerns regarding antibiotic resistance. (Note that bacteria in some of the manure samples exhibited resistance to flouroquinone — the antibiotic detected in the surface water sample.)

With respect to environmental impacts, it should also be noted that antibiotic residues can be photodegraded (by sunlight, for example) or by binding with organics (Karpati and Rubin, 1998). Like pathogens, whether or not these materials impact the environment may depend upon not only their quantity, but also how long they remain in the environment before they are inactivated.

Much more research is needed in this area to truly define the risks associated with antibiotic residues in the environment.

4.6.3 Herbicides

As noted in Chapter 2, clopyralid and picloram have been found in some yard waste composts in Washington, Pennsylvania, and New Zealand (Bezdicek et al., 2001). The contamination of these products was revealed when their use produced phytotoxic effects. Clopyralid is used on lawns and grass hay, whereas picloram is generally used to control weeds in pasture, rangeland, and industrial applications. At a composting facility in Spokane, WA, clopyralid was traced to lawn clippings. Both herbicides were introduced into a Washington State University (WSU) compost in Pullman, WA, through hay bedding and manure used as feedstock. (The hay had been sprayed with picloram and another herbicide.)

Bezdicek et al. (2001) report that producers have taken several measures to eliminate the inadvertent inclusion of pesticides in composts. One manufacturer temporarily discontinued the selling of a clopyralid product in Spokane when requested to do so, and the city of Spokane has increased public education campaigns to ensure that herbicide-treated grass is not composted. At WSU, composters now require straw and hay suppliers to certify that their products are free from herbicide contamination.

4.6.4 Endocrine Disruptors

As noted in Chapter 2, brominated diphenyl ethers (BDEs) have been identified in biosolids and endocrine disruptors have been found in both biosolids and manures. The relative contribution of BDEs from biosolids to the environment is unknown, and risks are uncertain, as well.

With respect to surfactants, a comprehensive literature review performed for the Water Environment Association of Ontario (WEAO, 2001) indicates that environmental impacts of these compounds in land applied-materials may be minimal (WEAO, 2001). This conclusion is based in part on studies of alkylphenol degradation in soils. At least one study indicates that aerobic conditions after biosolids land application can help to degrade these compounds. Topp and Starratt (2000) found that one form of these compounds, 4- nonyphenol, should degrade in well-aerated, arable soils. Hani (1990) also found that 4-nonyphenol degrades rapidly in soils. Further, literature indicates that alkylphnenols in land-applied biosolids (even immediately after application) should not impact plant growth as plant uptake is minimal, the compounds do not appear to leach to groundwater, nor is there transfer via the food chain to animals (WEAO, 2001).

With respect to estrogens, both natural and synthetic, laboratory research indicates that some common estrogens biodegrade rapidly in aerated soils at temperature and moisture conditions typical of growing seasons (Colucci, Bork, and Topp, 2001; Colucci and Topp, 2001). The researchers for these projects assessed the fate of the natural estrogens 17β-estradiol and estrone, and the xenoestrogen (synthetic estrogen) 17α-ethynlestradiol, which is used in oral contraceptives. Both studies looked at the fate of these estrogens when applied to a loam, a sandy loam and a silt loam. In addition to the dissipation of the estrogen itself, the researchers also observed a significant reduction in estrogenicity during the study period. Estrogen concentrations used in the studies were 10 mg/kg, however, far higher than might be expected in soil amendments and, accordingly, a subsequent study was conducted to ascertain the fate of land-applied estrogens at concentrations 1,000 times lower than the previous studies and more representative of potential concentrations in manures or biosolids (Topp, 2002). This laboratory study, assessing the natural estrogen 17β-estradiol and the xenoestrogen 17α-ethynylestradiol, found that, in the absence of preferential or surface flow, the risk of contamination of water resources from the application of biosolids or manures to agricultural soils during a temperate growing season is likely to be low.

The potential transport of land-applied estrogens to water resources has also been investigated by others. Assessing groundwater below an area of intensive livestock production in northwest Arkansas, Peterson, Davis, and Orndoff (2000) found that 17β-estradiol was affecting ground water quality in local mantled karst aquifers (karst topography is characterized by channels, fissures, and other mechanisms of preferential flow). Looking at surface flows, Nichols, et al. (1997) found that runoff from poultry litter-amended fields contained 17β-estradiol and that the estrogen persisted for 7 days under field conditions. Their estimates of persistence were based upon the detection of 17β-estradiol in runoff from a second storm 7 days following the initial runoff event. Estrogen concentrations were much lower in runoff from the second storm, however. The authors also found that the addition of alum to the litter significantly reduced estrogen transport, and that the concentrations of 17β-estradiol correlated to runoff pH and Total Organic Carbon (TOC) content (increases in both of these parameters increases 17β-estradiol in runoff). Others found no reduction in 17β-estradiol runoff with the addition of alum, however (Busheé, Edwards, and Moore, 1998).

Determining the presence of endocrine disruptors in the environment is easier than determining the specific impact of these compounds on the environment, however. Both the USEPA (1997b) and the NRC (2000) note that correlating a specific compound with observed ecological effects may be difficult because wildlife are exposed to a variety of influences, both chemical and nonchemical. NRC further notes that limitations and uncertainties regarding data, and even fundamental questions about observed effects, complicate evaluations of how — and how much — endocrine disruptors affect the environment. Further research is required to better understand this issue.

4.7 Odors and Airborne Pathogens

As noted previously, little quantitative data are available on land application odors. Anecdotal and general information demonstrates, without question, that manure and biosolids applications can cause nuisance odors.

More recently, however, the potential health effects of odorous emissions from biosolid and manure land application sites have been called into question (Lewis et al., 2000; Schiffman et al., 2000). Schiffman et al. (2000) noted that the most frequently reported health symptoms include eye, nose and throat irritation, headache, nausea, diarrhea, hoarseness, sore throat, cough, chest tightness, nasal congestion, palpitations, shortness of breath, stress, drowsiness, and alterations of mood. The authors further noted that, generally, these symptoms are transient, but that they may persist for longer periods or aggravate existing medical conditions in sensitive individuals.

For biosolids, concerns regarding health impacts appear to focus on Class B products (Lewis et al., 2000). This focus is reflected in a recent document released by the National Institute of Occupational Safety and Health (NIOSH), the *Guidance for Potential Risks to Workers Exposed to Class B Biosolids* (NIOSH, 2002). The NIOSH Guidance, which explicitly addresses occupational exposure, recommends basic health and safety procedures, such as hand washing, immunizations, and appropriate protective gear to prevent health problems for workers handling and applying Class B biosolids. Additionally, it recommends the incorporation of Class B biosolids into the soil (this procedure is generally advised as a Best Management Practice, but is not explicitly required by 40 CFR 503) and other measures.

Workers at biosolids application sites, biosolids processing facilities, and confined livestock facilities are generally presumed to represent a "worst case" with respect to exposure to odorants including ammonia, hydrogen sulfide and other odorous compounds, as well as airborne pathogens and most studies of health effects have focused on these individuals.

Less research has been conducted to assess the potential health impacts to people living near land application sites. To the authors' knowledge, the only extensive epidemiological study investigating this issue found no link between biosolids and illness (Brown, 1985). In a multiyear study, Brown (1985) compared the health of people and animals on 47 farms where biosolids had been applied to 45 farms where no biosolids had been used. The 3-year study reflected a total study population of almost 300 people and found no significant differences in respiratory and other illnesses in the people or animals on the biosolids-amended farms and those where no biosolids had been applied.

Other rigorous studies by researchers at Texas A&M University have indicated that the risk of infection to the public from airborne pathogens is low. In a study of airborne pathogen transport at a West Texas application site, Pillai et al. (1996), found that the application of biosolids pose little risk of exposure to airborne bacteria. The researchers also found, however, that agitation of the sludge (for example, during spreader loading operations) could increase the number of airborne bacteria in the vicinity of these operations, potentially posing a greater risk to biosolid appliers. In a subsequent risk assessment (Dowd et al., 2000), researchers found that, in fact, the risk of viral and bacterial infections to biosolids appliers was 3% and 2%, respectively (assuming wind speeds of 4.5 mi/hr (2 m/s) and 1 hour of exposure), but that the risk to nearby communities (about 4 miles, or 64 km, away) would be minimal under these same conditions. The authors were very careful to state that, as appropriate for a risk assessment, the models that they used are likely to overstate the actual risk.

At least one other study has claimed a direct relationship between odors from Class B land application and public health. In investigations performed long after a death in New Hampshire that they attribute to a Class B application, Lewis et al. (2000) surmise that an odorous biosolids application was responsible for the death and other illnesses in the area. The study, however, did not follow rigorous scientific procedures, and its results should be viewed accordingly.

Nonetheless, the question that Lewis et al. (2000) posed (specifically, "Is there a causal relationship between odors and public health?") is a valid one that bears further investigation. As noted by Schiffman et al. (2000), more objective scientific data is required, however, to relate chemicals, particulates, and microbes in odors to objective measures of health symptoms. Toward this end, USEPA is now leading a multiagency research on physiological effects of odor, particulates, endotoxin, and pathogens in air from land-applied biosolids and manure.

4.8 Summary

As shown in this chapter, there are potential environmental concerns associated with the use of biosolids, manures and fertilizers. The range of environmental and health issues identified cover the entire "scale of risk" from perceived concern (for example, an issue that is a concern, but has not been demonstrated to be problem after years of practice) to large scale, demonstrated impacts. Here we summarize the concerns identified, but, critically, assess them within the framework defined by the following two questions:

- What is the magnitude of this issue?
- Can risks be managed?

4.8.1 Nitrogen

With respect to nitrogen, fertilizers have been identified as a primary agricultural material responsible for nitrate contamination of groundwater (Bouchard, Williams, and Surampalli, 1992), most especially in the Midwest. These authors cite overapplication as the primary basis for contamination and further identify two reasons for this practice:

1. Farmers lack information to accurately predict nitrogen needs.
2. "The economic cost of overfertilization is less than the economic cost of underfertilization."

Despite the large volume of manure applied to the land, mineral fertilizers remain the largest source of land-applied nitrogen, contributing over 11 million tons (10 million metric tons) annually (1995 value, from USDA, 1997). Manures are estimated to add less than 6 million tons (5.4 million metric tons) of nitrogen annually, with volatilization and denitrification further reducing this value. Only about 3 million tons (2.7 million metric tons) of biosolids are land-applied (USEPA, 1999a), and with nitrogen concentrations averaging less than 5%, the relative impact of these materials is minimal). Nitrogen is also supplied by atmospheric deposition, septic tanks, and other sources.

Fortunately, the risks of nitrogen over-application — for all materials — can be managed through the use of NMPs, a growing practice applied to each of the materials discussed here. NMPs account for soil conditions, cropping practices, and nutrient sources to better predict nitrogen application requirements that will optimize crop yields without adversely impacting the environment.

4.8.2 Phosphorus

Phosphorus has been implicated as a leading contributor to surface water pollution, most especially eutrophication. Again, each of the materials studied here contribute phosphorus to the environment. Estimated phosphate fertilizer use was 4.4 million tons (4.0 million metric tons) in the US in 1995 (USDA, 1997), while phosphate in applied manures is estimated to be about 3.6 million tons (3.3 million metric tons). As with nitrogen, the contribution of biosolids-supplied P is negligible compared to fertilizer or manure sources. Moreover, materials used in municipal wastewater treatment tend to reduce the availability of biosolids-supplied phosphorus.

Recognizing the need to manage agricultural P inputs, agricultural institutions are beginning to incorporate phosphorus management on a broad scale. For example, the Natural Resources Conservation Service (NRCS) recently revised Code 590 (NRCS, 2000), its nutrient management standard, to incorporate P management tools (such as the Phosphorus Index). The NRCS revisions are meant to address concerns regarding manure applications in particular, but are appropriate for any phosphorus (or other nutrient) source applied to the land.

4.8.3 Metals

Land-applied metals remain a concern for some researchers, though as shown in this chapter, the preponderance of evidence suggests that metals applied in biosolids, manures and fertilizers remain in the surface soil. Years of research at sites where biosolids (the most investigated material with respect to metals) have been used further suggest that, when used at agronomic rates, crop yields are not impaired, but are essentially equivalent to (or better than) yields from mineral-fertilized or manured fields. Many of the biosolids studies cited in this chapter also investigated fertilizer and/or manure impacts, and no metal-related environmental impacts were identified.

With respect to animal health and land-applied metals, concerns regarding the impact of biosolids on molybdenosis in ruminants have been raised and are being researched. This issue, however, provides a good example of risk management. Specifically, animals that exhibit symptoms of this disease (copper deficiency) can be given copper supplements to completely offset any adverse health effects. As reported in this chapter, while at least one study potentially found Cu:Mo ratios below 2:1 (which would cause molybdenosis), other studies have found no such problems.

It is difficult to predict how much of an individual metal is applied to U.S. soils, but Table 4-6 can provide some relative guidance. As indicated by the table, it appears that far more metals are applied to U.S. lands through manure applications than through biosolids or fertilizers (an exception is cadmium supplied in phosphates). We caution that, as noted in Chapter 2, the different application rates — per unit area — for the products listed must be considered as well when considering metal applications. For example, though phosphates have the highest overall concentration of cadmium, their contribution is generally spread across a larger application area than biosolids or manures, and so unit area contributions (though still larger than biosolids or manures) are smaller than indicated by this table. Comparisons between manures and biosolids are somewhat more valid, however, since the application rates for these products can be similar.

With respect to risks from land-applied metals, USEPA (2001d) noted that manures applied at agronomic rates "should pose little risk to human health and the environment. However, repeated applications of manures above agronomic rates could result in exceedances of the cumulative metal loading rates established in Part 503, thereby potentially impacting human health and the environment."

Like nutrients, metal concentrations in biosolids and manures can be managed to minimize any risks, however small. As noted in Chapter 2, biosolids metal concentrations continue to fall through continued implementation of industrial pretreatment programs. For manures, practitioners are researching ways to reduce manure metals through diet modifications. Monitoring of metal application rates — as is required for biosolids — further reduces risks associated with land-applied metals.

Risk assessments accepted by the mainstream of the scientific community have demonstrated that metals supplied through most fertilizers and biosolids do not pose a risk to human health when applied at recommended rates. A risk assessment addressing manure metals has not yet been performed, but based upon the information presented here, would be of interest. Although it is impossible to say what the outcome of such an assessment might be, it is critical to note that, like biosolids, manures have generally been applied for crops for many years without any observed impacts on human or plant health from metals that they contain.

Table 4-6
Estimated Total Metals Applied to Land from Various Products

Metal	Metal Concentration (Average), mg/dry kg				
	Biosolids(1)	Swine Manure(2)	Poultry Manure(3)	Phosphate Fertilizer (4)	Micronutrient Mix (5)
Arsenic	5.01	3.7	13	11.3	41.9
Cadmium	4.45	2.5	2.4	65	27.9
Chromium (5)	59.3	NA	NA	173	230
Copper	425	305	465	56.6	29,650
Lead	75.9	7.6	46	12.2	1798
Mercury	1.79	ND	NA	0.1	0.1
Molybdenum	12.1	NA	19	NA	NA
Nickel	33.2	29.3	16	27.5	12.5
Selenium (5)	3.8	NA	NA	NA	NA
Zinc	735	555	602	240.2	77,300

Metal	Applied Metals (tons/year) (1)				
	Biosolids (2)	Swine Manure (3)	Poultry Manure (4)	Phosphate Fertilizer (5)	Micronutrient Mix (5)
Arsenic	14.2	33.3	189	49.7	1.17
Cadmium	12.6	22.5	33	286	0.78
Chromium	168	NA	NA	761	6.44
Copper	1,202	2,745	6,743	249	830
Lead	215	68.4	667	53.7	50.3
Mercury	5.07	NA	NA	0.0	0.003
Molybdenum	34.2	NA	NA	NA	NA
Nickel	94.1	264	NA	121	0.35
Selenium	10.9	NA	NA	NA	NA
Zinc	2,081	4,995	8,729	1,057	2,164

Notes:

(1) Based on metal concentrations presented in Table 2-15.

(2) Assumed application of 2.83 million dry tons annually (USEPA, 1999a)

(3) Assumed application of 9 million dry tons annually (USEPA, 1998)

(4) Assumed application of 14.5 million dry tons annually (EPA, 1998)

(5) Assumed application of 4.4 million tons annually (USDA, 1997)

4.8.4 Pathogens

As reviewed in Chapter 2 and here, concentrations of bacteria used to indicate total pathogen content are higher in manures than in biosolids, and unlike biosolids, their concentrations in manures are not strictly regulated. Accordingly, there have been limited illness outbreaks associated with manure use, but no demonstrated impacts of a similar nature from biosolid use.

Considering that more than 150 million dry tons of manure is applied to the land annually, the number of known illness outbreaks is exceedingly small. Nonetheless, the adoption of simple management practices (such as setbacks and filter strips — which are applicable for nutrient management as well) might warrant consideration to reduce any potential risks associated with manure pathogens.

4.8.5 Other Chemicals

Primary considerations at this time include dioxin, which is slated to be regulated in biosolids under proposed revisions in 40 CFR 503 and endocrine disruptors, which are contained in biosolids and manures. Like the 503 metal limits, the proposed dioxin limit was developed through a risk assessment and it is expected that the vast majority of biosolids generated will be able to meet the proposed limit. Dioxin limits have not been established for fertilizers, although high dioxin concentrations have been found in some micronutrients made from recycled hazardous wastes (see Chapter 2); these products do not represent the bulk of fertilizers generated, however.

Little is currently known about direct impacts of biosolids or manure-supplied endocrine disruptors on the environment, although it appears that synthetic and natural estrogens and some surfactants degrade readily in aerobic soils (Colucci, Bork, and Topp, 2001; Colucci and Topp, 2001; Topp and Starratt, 2000; Hani, 1990). Additional study is required in this area however, as this issue is complex, with multiple stressors potentially acting in concert to affect a given impact. Trying to determine the specific effect of a single compound, or even several related compounds, is not a simple undertaking and extensive research may be required to address this issue.

4.8.6 Odors and Other Emissions

Odors are an acknowledged part of biosolids and manure land applications. To date, odor has primarily been viewed as a nuisance issue, but increasingly, the health impacts of odors from land application sites, as well as other application emissions (such as bacteria) are now being questioned, despite the fact that no documented health impacts have been identified. As reviewed here, studies have shown that airborne pathogens from Class B biosolids are a potential risk for workers at biosolids application sites (Dowd et al., 2000), although the risk was small and potentially overestimated. No similar studies have been performed for manures, but based upon the higher indicator bacteria concentrations in manures, the results of such studies would be of interest.

4.9 Concluding Remarks

The above discussions indicate that the relative risk (based upon contributions to the environment) from amendments and fertilizers varies by parameter. For example, improperly applied mineral fertilizers may be the greatest concern with respect to nitrogen impacts, but manures may be a greater concern with respect to phosphorus, some metals, and pathogens.

Interestingly, concerns regarding metals and pathogen risks have focused until recently on biosolids. Based on research presented here, however, it appears that the risks associated with biosolids use are minimal and no greater than risks associated with manure use. Thus, the attention directed toward biosolids and concerns regarding their use may be disproportionate, especially considering the fact that these stringently regulated materials are applied to less than 1% of the agricultural land in this nation (Draeger et al., 1999).

This study's findings regarding the minimal risk of biosolids applications supports an earlier NRC assessment (NRC, 1996) which found that:

> "...society produces large volumes of treated municipal wastewater and sewage sludge that must be either disposed of or reused. While no disposal or reuse option can guarantee complete safety, the use of these materials in the production of crops for human consumption, when practiced in accordance with existing federal guidelines and regulations, presents negligible risk to the consumer, to crop production, and to the environment."

More recently, an NRC study found no documented health impacts from biosolids land application (NRC, 2002).

When making decisions about whether to use the materials here, it is important to remember that known risks can be managed through proper application.

For all materials, some questions remain. Ecosystem impacts have not been assessed in detail. For amendments in particular, the impact of endocrine disruptors and health impacts from odors and other emissions require additional study. For biosolids, a new risk assessment reflecting advances in risk assessment science over the last decades has been recommended by the NRC (NRC, 2002).

In sum, however, it appears that known risks from each of the materials studied can be managed. Moreover, these manageable risks must be carefully weighed against the considerable benefits provided by the land application of amendments and fertilizers.

CHAPTER 5.0

RESEARCH NEEDS

On the basis of this study, the project team identified the following research needs:

Manure metals: Metal concentrations in manures have not been heavily researched and data is limited, especially for non-nutritive elements. As shown in Chapters 2 and 4, limited data indicate that metal concentrations in these materials are comparable to biosolids. The development of a larger database, for 503 metals in particular, is recommended to further assess the potential impact of manure-applied metals.

An assessment of metals availability in these materials — through laboratory and field studies — may also be warranted if the database confirms that metal concentrations are comparable to biosolids, especially considering the volume of manures applied to the land.

Amendment pathogens: The database for 503 indicators of pathogens in biosolids is large and does not need to be supplemented. Fecal coliforms and *Salmonella* concentrations in manures have been assessed to some extent, but that database is relatively small and should be expanded further, as an indicator of pathogen content.

Other pathogens implicated in outbreaks of various illnesses — specifically emerging pathogens such as *E. coli* O157:H7, *Giardia,* and *Crytosporidium* — have not been studied to a great extent in biosolids or manures. As we develop the capability to more easily measure low concentrations of emerging pathogens, direct measurement of pathogenic organisms should increase our understanding of both the prevalence and potential impact of these organisms. (WERF, USDA, and USEPA are currently leading research efforts in this area.)

In sum, we recommend further assessments of emerging pathogens in biosolids and manures, and fecal coliforms and *Salmonella* testing for manures only.

For both biosolids and manures, the fate and transport of pathogens in the environment need to be better understood. To thoroughly understand potential risks, however, it is critical that pathogen transport from both products be assessed, especially considering the higher concentrations of indicator bacteria in manures and the larger quantities of these materials that are land applied.

Long-term field studies of all materials (at sites across the country) under documented and proper agronomic management: As noted in this report, many academic studies investigating biosolids impacts have used application rates far exceeding agronomic rates and/or metals application rates that would not be allowed under today's 503 rule. Although these studies can provide a worst-case perspective, they do not necessarily reflect the impacts of properly applied biosolids and, particularly, should not be used to assess the relative impacts of biosolids, manures, and chemical fertilizers. Moreover, programs such as the National Biosolids Partnership's Environmental Management System (EMS) may take biosolids management to a new level, improving biosolids quality and field management practices further.

For these reasons, we strongly advocate new long-term field studies preferably working with programs that have implemented an EMS and/or continued monitoring of programs documented in WERF's 1993 report *Document Long-Term Experience of Biosolids Land Application Programs* (Jacobs et al. 1993), as well as comparative studies at well controlled study sites such as the Rosemount Experiment Station in Minnesota.

Improved model for the economic assessment of organics applications: To date, most studies have focused on the fertilizer replacement value of soil amendments, ignoring or poorly reflecting the indirect economic benefits of improved soil tilth, reduced compaction, reduced erosion, and other advantages from amendment use. Accordingly, we recommend the development of a more holistic approach to estimating the economic benefit derived from the use of these materials.

Comprehensive ecological assessment for soil amendments and chemical fertilizers: As noted earlier, an ecological assessment was conducted for biosolids, but it was limited in scope and major conclusions of the report — particularly those regarding nutrients — are potentially applicable to manures and fertilizers as well. Accordingly, we recommend an additional study of all materials applied at typical rates to realistically assess the impacts of their use on the ecosystem.

Endocrine disruptor and pharmaceutical content in amendments and the relative impact of these compounds on the environment: Research into the concentrations of endocrine disruptors and pharmaceuticals is beginning but must be greatly expanded to provide a reasonable database for further assessment. Additionally, the fate and relative impact of these compounds — compared to other sources in the environment — must be carefully assessed.

Odor impacts on health: Concerns have been raised regarding the health impacts of odors from manure and biosolids land application. Further research into the relationship between odors and health is required to ensure that management programs adequately address this concern.

Manure nutrient availability: Uncertainties regarding nitrogen availability, in particular, can lead to over or under-application of this nutrient. The development of a simple test to more accurately estimate plant available nitrogen (PAN) in manures would benefit appliers of this material and, potentially, the environment as well. Ongoing research in this area will hopefully yield results in the next several years.

Overall, we must stress that any investigations into both the benefits and risks of the materials studied here must be rigorously documented and balanced, reflecting the *relative* benefits and risks compared to other materials that are used in similar ways.

REFERENCES

Abd El-Hay, G.H., and J.S. Angle. (1994) Sludge-borne salt effects on soybean growth. In: *Sewage Sludge: Land Utilization and the Environment.* Edited by C.E. Clapp, W.E. Larson, and R.H. Dowdy. SSSA Miscellaneous Publication.

Agency of Toxic Substances and Disease Registry. (1994) Biological indicators of exposure to cadmium an lead, Palmerton, PA, Part II. U.S. Department of Health Services. Atlanta, Georgia.

Alexander, R., and N. Goldstein. (1999) Setting metals standards for fertilizer. *BioCycle* **40(5)**: 66-69.

Allmaras, R., Schomberg, H., Douglas, C., and T. Dao. (2000) Soil organic carbon sequestration potential of adopting conservation tillage in U.S. croplands. *Journal of Soil and Water Conservation.* **55**: 365-373.

American Society of Agricultural Engineers. (2000) *ASAE Standards 2000* St. Joseph, Michigan: American Society of Agricultural Engineers.

Andres, S.A. (1995) Nitrate loss via ground water flow, coastal Sussex County, Delaware. In: *Animal Waste and the Land-Water Interface*. Edited by Kenneth Steele. CRC Press. Boca Raton, FL.

Association of American Plant Food Control Officials (2001) *SUIP 25 Metals in Fertilizers (proposed)*. Available at: www.aapfco.org/SUIP25MidYr01.htm.

Bali, K.M., Escobosa, I.G., Guerrero, J.N., Crohn, D.M., and M.E. Grismer. (1998) Effect of biosolids on infiltration rate in clay soils. Written for presentation at the *1998 American Society of Agricultural Engineers Annual International Meeting.*

Barker, A.B., Stratton, M.L., and J.E. Rechcigl. (2000) Soil and by-product characteristics that impact the beneficial use of by-products. In: *SSSA Book Series: 6: Land Application of Agricultural, Industrial, and Municipal By-Products*, edited by J.F. Power and W.A. Dick.

Barbarick, K.A., Ippolito, J.A., and D.G. Westfall. (1998) Extractable trace elements in the soil profile after years of biosolids application. *J. Environ. Qual.* **27:**801-805.

Baveye, P.M., McBride, M.B., Boudin, M.B., Hinesly, D., Dahdoh, T.D. and M.R. Abdelsabour. (1999) Mass balance and distribution of sludge-borne trace elements in a silt loam soil following long-term application of sewage sludge. *Science of the Total Environment.* **227:**13-28.

Bennett, G.F., Gopalan, P., and A. Senjupta. (1989) Effects of cement-kiln dust on the mobility of heavy metals in treatment of wastewater treatment plant sludge. A report for Thomas Edison Program, The Edison Seed Development Fund, State of Ohio, Columbus, Ohio.

Bergström, L.F., and H. Kirchmann. (1999) Leaching of total nitrogen-15-labeled poultry manure and inorganic nitrogen fertilizer. *J. Environ. Qual.*. **28**:1283-1290.

Bertrand, J.D., Lutrick, M.C., Edds, G.T., and R.L. West. (1981) Metal residues in tissue, animal performance, and carcass quality with beef steers grazing on Pensacola Bahiagrass pastures treated with liquid digested sludge. *J. Animal Sci.* **53:**146-163.

Berti, W.R., and L.W. Jacobs. (1996) Chemistry and phytotoxicity of soil trace elements from repeated sewage sludge applications. *J. Environ. Qual.* **25:**1025-1032.

Bevacqua, R.F., and V.J. Mellano. (1994) Cumulative effects of sludge compost on crop yields and soil properties. *Common. Soil Sci. Plant. Anal.* **25**(3&4): 39-406.

Bezdicek, D., Fauci, M., Caldwell, D., Finch, R., and J. Lang. (2001) Persistent herbicides in compost. *BioCycle.* **42(7)**: 25-23.

Bouchard, D.C., Williams, M.K. and R.Y. Surampalli. (1992). Nitrate contamination of groundwater: sources and potential health effects. Journal AWWA. September: 85-90.

Bouldin, D.R. (1997) Why guidelines for beneficial use of sludges in agriculture are different and estimates of alternatives. Soil Science Society of North Carolina Proceedings, Vol. XL. Available at http://www.cfe.cornell.edu/wmi/Sludge/WhyGuide/WhyGuide.html.

Bray, B.J., Dowdy, R.H., Goodrich, R.D. and D.E. Pamp. (1985) Trace accumulations in tissues of goats fed silage produced on sewage sludge-amended soil. *J. Environ. Qual*. **14**: 114-118.

Brinton, W.F., Trankner, A. and M. Droffner. (1996) Investigations into liquid compost extracts. *BioCycle*. **37(11)**: 68-70.

Brown, R.E. (1985) Demonstration of acceptable systems for land disposal of sewage sludge. Water Engineering Research Lab. EPA 600/Z-85-062. Cincinnati, Ohio.

Burnham, J.C. (1999) *The Unity cross-reactor process*. Demonstration Project at Cypress Chemical Company. Helena, Arkansas.

Busheé, E.L., Edwards, D.R., and P.A. Moore, Jr. (1998) Quality of runoff from plots treated with municipal sludge and horse bedding. *Transactions of the ASAE*. **41(4)**: 1035-1041.

Byers, D.E., and R.L. Mikkelsen (1999) Ammonia volatilization from swine waste following application to cropland. *Proceedings from 1999 NC State University Animal Waste Management Symposium*. North Carolina State University, North Carolina.

Camobreco, V.J., Richards, B.K., Steenhuis, T.S., Peverly, J.H., and M.B. McBride (1996) Movement of heavy metals through undisturbed and homogenized soil columns. *Soil Science.* **161:** 740-750.

Camp Dresser and McKee (1998) *Biosolid Utilization Evaluation Program Final Project Report.* Prepared for Connecticut Department of Environmental Protection.

Campagnolo, E.R., and C.S. Rubin (1998) *Report to the state of Iowa Department of Public Health on the investigation of chemical and microbial constituents of ground and surface water proximal to large-scale swine operations.* Follow-up Investigation to the Centers for Disease Control and Prevention Workshop.

Caspers-Simmet, J. (2000) What Are Credits Worth? Depends on U.S. Commitment to Clean Air. Agri News. Available at: http://webstar.agrinews.com/agrinews/287531819483348.bsp.

Centers for Disease Control and Prevention (1996) Spontaneous abortions possibly related to ingestion of nitrate contaminated well water. *Morbidity and Mortality Weekly Report.* 45:569-572.

Chaney, R.L., Stoewsand, G.S., Furr, A.K., Bache, C.A., and D.J. Lisk. (1978a) Elemental content of tissues of guinea pigs fed Swiss chard grown on municipal sewage sludge-amended soil. *J. Agr. Food Chem.* 26:994-995.

Chaney, R.L., Stoewsand, G.S., Furr, A.K., Bache, C.A., and D.J. Lisk. (1978b) Cadmium deposition and hepatic microsomal induction in mice fed lettuce grown on municipal sludge-mended soil. *J. Agric. Food Chem.* 26:992-994.

Chaney, R.L. (1994) Trace metal movement: soil-plant systems and bioavailability of biosolids-applied metals. In: *Sewage Sludge: Land Utilization and the Environment.* Edited by C.E. Clapp, W.E. Larson, and R.H. Dowdy. SSSA Miscellaneous Publication.

Chang, C., Sommerfeldt, T.G., and T. Entz. (1990) Rates of soil chemical changes with eleven annual applications of cattle feedlot manure. *Can. J. Soil Sci.* **70**: 673-681

Chang, C., and T. Entz. (1996) Nitrate leaching losses under repeated cattle feedlot manure applications in southern Alberta. *J. Environ. Qual.* **25**:145-153.

Christie, P. and J.A.M. Beattie. (1989) Grassland soil microbial biomass and accumulation of potentially toxic metals from long-term slurry application. *J. Applied Ecology.* **26**:597.

Clapp, C.E., Dowdy, R.H., Linden, D.R., Larson, W.E., Hormann, C.M., Smith K.E., Halbach, T.R., Cheng, H.H., and R.C. Polta. (1994) Crop yields, nutrient uptake, soil and water quality during 20 years on the Rosemount sewage sludge wastewater. In: *Sewage Sludge: Land Utilization and the Environment.* Edited by C.E. Clapp, W.E. Larson, and R.H. Dowdy. SSSA Miscellaneous Publication.

Colluci, M.S., Bork, H. and E. Topp. (2001) Persistence of estrogenic hormones in agricultural soils I: 17β-estradiol and estrone. *J. Environ. Qual.* **30**: 2070-2076

Colluci, M.S and E. Topp. (2001) Persistence of estrogenic hormones in agricultural soils II: 17α-ethynylestradiol. *J. Environ. Qual.* **30**: 2077-2080

Cook, J. and J. Beyea (1995). Potential toxic and carcinogenic chemical contaminants in source separated municipal solid waste composts: review of available data and priorities for action. US Composting Council, Alexandria, VA.

Council for Agricultural Science and Technology (1995) *Waste Management and Utilization in Food Production and Processing.* Task Force Report (No. 124) Ames, IA.

Council for Agricultural Science and Technology (1996) *Integrated Animal Waste Management.* Task Force Report (No. 128) Ames, IA.

Daniel, T.C., Edwards, D.R., and D.J. Nichols. (1995) Edge-of-field losses of surface-applied animal manure. In: *Animal Waste and the Land-Water Interface.* Edited by Kenneth Steele. CRC Press. Boca Raton, FL.

Dao, T.H. (1999) Coamendments to modify phosphorus extractability and nitrogen/phosphorus ratio in feedlot manure and composted manure. *J. Environ. Qual.* **28**:1114-1121.

Dao, T. H., Sikora, L. J., Hamasaki, A. and R. L. Chaney (2001) Manure phosphorus extractability as affected by aluminum- and iron by-products and aerobic composting. *J. Environ. Qual.* **30**:1693-1698.

Dentel, S.K., Allen, H.E., Srinivasarao, C., and J. Divincenzo. (1993). Effects of surfactants on sludge dewatering and pollutant fate. Third Year Completion Report Project No. 6, Prepared for Water Resources Center, University of Delaware.

Dickerson, G.W. (1996) Compost dressing helps chili peppers. *Biocycle.* **37(3)**: 80-82

Dickerson, G.W. (1999) Description and Use of Municipal Solid Waste Composts in New Mexico. New Mexico State University, College of Agriculture and Home Economics Publication No. 562

Dowd, S.E., Gerba, C.P., Pepper, I.L., and S.D. Pillai (2000). Bioaerosol transport modeling and risk assessment in relation to biosolid placement. *J. Environ Qual.* **29**:343-348.

Dowdy, R.H., Bray, B.J., Goodrich, R.D., Marten, G.C., Pamp, D.E., and W.E. Larsen (1983) Performance of goats and lambs fed corn silage on sludge-amended soil. *J. Environ. Qual.* **12**:467-472.

Dowdy, R.H., Clapp, C.E., Linden, D.R., Larson, W.E., Halbach, T.R., and R.C. Polta (1994) Twenty Years of Trace Metal Partitioning on the Rosemount Sewage Sludge Watershed. In: *Sewage Sludge: Land Utilization and the Environment.* Edited by C.E. Clapp, W.E. Larson, and R.H. Dowdy. SSSA Miscellaneous Publication.

Draeger, K., Pundsack, J., Jorgenson, M., and P. Mulloy (1999) *Watershed Efffects of Biosolids Land Application: Literature Review.* Water Environment Research Foundation. Project 96-REM-2.

Dressler, R.L., Storm, G. L., Tzilkowski, W.M., and W.E. Sopper (1986) Heavy metals in cottontail rabbits on mined lands treated with sewage sludge. J. Environ. Qual. 15:278-281.

Drury, D.R. and A. Lloyd. (1997) Cryptosporidium – a regulators viewpoint. In: 1997 International symposium on waterborne Cryptosporidium proceedings. Edited by C.R. Fricker. American Water Works Association. Denver, CO. 369-375.

Efroymson, R.A., Sample, B.E., Luxmoore, R.J., Tharp, M.L, L.W. Barnthouse. (1998) *Evaluation of the Ecological Risks Associated with Land Application of Municipal Sewage Sludge.* Oak Ridge National Laboratory, Oak Ridge TN.

Eghball, B., Binford, G.D., and D.D. Baltensperger. (1996) Phosphorous movement and adsorption in a soil receiving long-term manure and fertilizer application. *J. Environ. Qual.* **25**:1339-1343.

Epstein, E. (1975) Effect of sewage sludge on some soil physical properties. *J. Environ. Qual.* **4**:139-142.

Epstein, E., Taylor, J.M., and R.L. Chaney. (1976) Effects of sewage sludge compost applied to soil on some physical and chemical properties. *J. Eviron. Qual.* **5**: 422-426.

Epstein, E. and J.F. Donovan. (1992). Pathogens in compost and their fate. In: *Pathogens in Sludge: What Does it Mean?* Water Environment Federation Pre-Conference Seminar, New Orleans, LA, July 1992.

Epstein, E., Chaney, R.L., Henry, C., and T.J. Logan. (1992) Trace elements in municipal solid waste compost. *Biomass and Bioenergy.* **3**:3-4.

Epstein, E. (1997). *The Science of Composting.* Lancaster, PA. Technomic Publishing Company, Inc.

Erdat, O., and W.B. Voorhees. (1990) Economic consequences of soil compaction." ASAE Paper No. 90-1089. St. Joseph, MI: American Society for Agricultural Engineers.

Farrell, M. (1996) Compost pays off in the orchard. *Biocycle.* **37(10)**: 40-42.

Faust, J. G. and R. L. Oberst. (1996) Economic value of biosolids to farmers. *BioCycle.* **37** (1): 67-69.

Fertilizer Institute. (1982) *The Fertilizer Handbook.* Washington, DC.

Fleming, R., McLellan, J., Alves, D., Hilborn, D., Pintar, K., and M. MacAlpine. (1997) *Cryptosporidium* in livestock, manure storages, and surface waters in Ontario. *Ontario Farm Environmental Coalition Ontario Ministry of Agriculture, Food and Rural Affairs and Agri-Food Canada*. Toronto, Ontario.

Food and Agricultural Organization of the U.N. (1999) Poor management of chemical inputs. Available at: http://www.fao.org

Foster Wheeler (1998) *Development of Risk Based Concentrations for Arsenic, Cadmium and Lead in Inorganic Commercial Fertilizers.* Prepared for California Department of Food and Agriculture.

Gadjos, R. (1997) Effects of two composts and seven commercial cultivation media on germination and yield. *Compost Science and Utilization.* **5(1):**16-37

Garcia, A., and A.S. Mangaroo. (1994) Residual concentration of selected heavy metals in a sewage sludge-amended soil and uptake by costal bermudagrass. In: *Sewage Sludge: Land Utilization and the Environment.* Edited by C.E. Clapp, W.E. Larson, and R.H. Dowdy. SSSA Miscellaneous Publication.

Gallimore, L.E., Basta, N.T., Storm, D.E., Payton, M.E., Huhnke, R.H., and M.D. Smolen. (1999) Water treatment residual to reduce nutrients in surface runoff from agricultural land. *J. Environ. Qual.* **28:**1474-1478.

Giddens, J., and A.P. Barnett. (1980) Soil loss and microbial quality of runoff from land treated with poultry litter. *J. Environ. Qual.* **9**:518.

Gilmour, J.T., Cogger, C.G., Jacobs, L.W., Wilson, S.A., Evanylo, G.K, and D.M. Sullivan. (2000) *Estimating Plant-available Nitrogen in Biosolids.* Water Environment Research Foundation. Project 97-REM-3.

Giusquiani, P.L., Pagliali, M., Gigliotti, G., Businelli, D., and A. Benetti. (1995) Urban waste compost: effects on physical, chemical and biological soil properties. *J. Environ. Qual.* **24**:175-182.

Glenn, J., and D. Block. (1999) MSW Composting in the United States. *BioCycle* **40(11):**42-48.

Goldstein, J. (1998) Compost suppresses disease in the labs and on the fields. *Biocycle.* **39(11):** 62-64.

Goldstein, N. (2000) The state of garbage. *BioCycle* **41(4)**: 32-39.

Golueke, *C.G. (1983) Epidemiological aspects of sludge handling and management. Part II. Biocycle*. **24(3):** 52-58

Golueke, C.G., Diaz, L.F., and S. Gurkewitz (1989) Technical analysis of multi-compost products. *Biocycle.* **30(6)**: 55-57.

Goolsby, D.A., and W.A. Battaglin. (1997) Sources and flux of nitrate in the Mississippi River system. Presented at the American Geophysical Union Meeting. Baltimore, MD.

Goss, M.J., Rollins, K.S., McEwan, K, Shaw, J.R. and H. Lammers-Helps. (2002). The Management of Manure in Ontario with Respect to Water Quality. Commissioned Paper No. 6 for the Walkerton Inquiry. Available at: www.walkertoninquiry.com

Gouin, F.R. (1997) It's good for what ails you. *Biocycle.* **38(9):**61-62.

Grace, N., Carr, G., and C.S. Fox. (1994) Characteristics of pelletized products derived from the direct heat drying of three types of biosolids: a comparative evaluation with end user market requirements. Proceedings of the Water Environment Federation 67th Annual Conference and Exposition. Chicago, IL.

Grattan, L.M., Oldach, D., Perl, T.M., Lowitt, M.H., Matuszak, D.L., Dickson, C., Parrott, C., Shoemaker, R.C., Kauffman, C.L., Wasserman, M.P., Hebel, J.R., Charache, P., J.G. Morris, Jr. (1998) Learning and memory difficulties after environmental exposure to waterways containing toxin-producing Pfiesteria or Pfiesteria-like dinoflagellates. *Lancet* **352**: 53239.

Hancock, D.D., Besser, T.E., Kinsel, M.L., Tarr, P.I., Rice, D.H. and M.G. Paros. (1994) The prevalence of *Escherichia coli* O157:H7 in dairy and beef cattle in Washington state. *Epidemiology and Infection.* 113: 199-207.

Hale, R.C., La Guardia, M.J., Harvey, E.P., Gaylor, M.O., Mainor, T.M., and W.H. Duff. (2001) Persistent pollutants in land-applied sludges. *Nature.* **412:**140-1.

Hani, H. (1990) The analysis of inorganic and organic pollutants in soil with special regard to their bioavailability. *Int. J. Environ. Anal. Chem.* **39**: 197-208.

Hansen, L.G., and T.D. Hinesly. (1979) Cadmium from soil amended with sewage sludge: effects and residues in swine. *Environ. Health Perspect.* **28:**51-57.

Hardwick, D.C. (1995) *Agricultural problems related to odor prevention and control. Odor Prevention and Control of Organic Sludge and Livestock Farming.* Edited by V.C. Nielsen, J. H. Voorburg, and P. L'Hermite. Elsevier Applied Science Publishers, NY.

He, X.T., Logan, T.J. and S.J. Traina. (1995) Physical and chemical characteristics of selected municipal solid waste composts. *J. Environ. Qual.* **24**: 543-552.

Heckman, J.R., Angle, J.S., and R.L. Chaney. (1987) Residual effects of sewage sludge on soybean: I. Accumulation of soil and symbiotically fixed nitrogen. J. Environ. Qual. **16(2):**118-124.

Henry, C.L., and R.B. Harrison. (1992) Comparing yard waste and sludge compost. *Biocycle.* **33(2)**:42-47.

Hoitink, H., Shang, W., Han, D., and W. Dick. (1997) Making compost to suppress plant disease. *BioCycle*. **38(4):** 40-42.

Holmgren, G., Meyer, M., Chaney, R., and R. Daniels. (1993) Cadmium, lead, zinc, copper, and nickel in agricultural soils of the United States of America. *J. Environ. Qual.* **22**: 335-348.

Hormann, C.M., Clapp, C.E., Dowdy, R.H., Larson, W.E., Duncomb, D.R., Halbach, T.R. and R.C. Polta. (1994) Effect of lime-cake municipal sewage sludge on corn yield, nutrient uptake, and soil analyses. In: *Sewage Sludge: Land Utilization and the Environment.* Edited by C.E. Clapp, W.E. Larson, and R.H. Dowdy. SSSA Miscellaneous Publication.

Humpert, C.P. (2000) New trends in sustainable farming build compost use. *BioCycle* **41(7):** 30-33.

Hussong, D., Burge, W.D. and N.K. Enkiri. (1985) Occurrence, growth, and suppression of salmonellae in composted sewage sludge. *Applied and Environmental Microbiology* **50**:887-893.

Ibekwe, A. M., Angle, J.S., Chaney, R.L. and P. van Berkum. (1995). Sewage sludge and heavy metal effects on nodulation and nitrogen fixation of legumes. *J. Environ. Qual.* **24**:1199-1204.

Ibekwe, A. M., Angle, J.S.,Chaney, R.L. and P. van Berkum. (1998) Zinc and cadmium effects on rhizobhia and white clover using chelator-buffered nutrient solution. *Soil Sci. Soc. Am. J.* **62**:204-211.

Ireland, S.J. and M.W. Thayer (2000) *Process development: biosolids enrichment from pilot to full-scale demonstration.* Proceedings of the 14th Annual Residuals and Biosolids Conference of the Water Environment Federation. Boston, MA February 27- March 1, 2000.

Jacobs, L.W., Carr, S., Böhm, S., and J.R. Stukenberg. (1993) *Document Long-Term Experience of Biosolids Land Application Programs.* Water Environment Research Foundation. Project 91-ISP-4.

Jacobson, L., Schmidt, D., Nicolai, R., and J. Bicudo. (1998) Odor control for animal agriculture. BAEU-17. University of Minnesota Biosystems and Agricultural Engineering. Available at: www.bae.umn.edu/extens/aeu/baeu17.html.

Jinadasa, K.B.P.N., Milham, P.J., Hawkins, P.S., Cornish, P.S., Williams, P.A., Kaldor, C.J., and J.P. Conroy. (1997). Survey of cadmium levels in vegetables and soils of greater Sydney, Australia. *J. Environ. Qual.* **26**: 924-933.

Johnston, A.E., McGrath, S.P., Poulton, P.R., and P.W. Lane. (1989) Accumulation and loss of nitrogen from manure, sludge and compost: Long-term experiments at Rothamsted and Woburn, Chap. 10 in *Nitrogen in Organic Wastes Applied to Soils*. Edited by J.A. Hansen, and K. Henriksen. Academic Press, NY.

Jokela, W.E. (1992) Nitrogen fertilizer and dairy manure effects on corn yield and soil nitrate. *Soil Sci. Soc. Am. J.* **56:**148-154.

Jones, K.C., Symon, C.J., and A.E. Johnston. (1987) Retrospective analysis of an archived soil collection. II. Cadmium. *Sci. Total Environ.* **67:**75-89.

Jones, K.C. and A.P Sewart (1997) Dioxins and furans in sewage effluent sludges: a review of their occurrence and sources in sludge and of their environmental fate, behavior, and significance in sludge-amended agricultural systems. *Critical Reviews in Environmental Science and Technology*. **27**:1-85.

Jones, P.W. (1980) Health hazards associated with the handling of animal wastes. *Vet. Rec.* 106: 4-7.

Jones, P.W., Collins, P., Brown, G.T.H., and M. Aitken. (1982) Transmission of *Salmonella mbandaka* to cattle from contaminated feed. *J. Hyg. Camb.* **88**: 255-263.

Kashmanian, R.M., Kluchinski, D., Richard, T.L, and J.M. Walker. (2000) Qualities, characteristics, barriers, and incentives for use of organic municipal by-products. In: *SSSA Book Series: 6: Land Application of Agricultural, Industrial, and Municipal By-Products*, edited by J.F. Power and W.A. Dick.

Karpati, A. and C. Rubin. (1998) Report of a pilot environmental investigation around large poultry operations in Ohio. Centers for Disease Control and Prevention.

Kellog, R.L., Lander, C.H., Moffitt, D.C., and N. Gollehon. (2000) Manure Nutrients Relative to the Capacity of Cropland and Pastureland to Assimilate Nutrients: Spatial and Temporal Trends for the United States. USDA-NRCS Publication No. nps00-0579. Available at: http://www.nrcs.usda.gov/technical/land/pubs/manntr.pdf.

Khaleel, R., Reddy, K.R., and M.R. Overcash. (1981) Changes in soil physical properties due to organic waste applications: a review. *J. Environ. Qual.* **2:**133-141.

Kintzer, B. (2002) Personal Communication.

Kostewicz, S.R. (1993) Pole bean yield as influenced by composted yard waste soil amendment. *Proc. Fla. State Hort. Soc.* 106:1993.

Landers, C.H. Moffitt, D., and K. Alt. (1998) Nutrients available from livestock manure relative to crop growth requirements. *Resource Assessment and Strategic Planning Working Paper 98-1.*

Landers, J. (2000) Dioxin monitoring requirements for land-applied biosolids proposed by EPA. *Biosolids Technical Bulletin.* **1**: 1-2.

Lazarovits, G. and K. Conn. (1997) Assessment of the influence of manures for the control of soil borne pests including fungi, bacteria, and nematodes. COESA Report No.: RES/MAN-010/97. Research Branch, Agriculture and Agri-Food Canada.

Linden, D.R., Larson, W.E., Dowdy, R.H., and C.E. Clapp (1995) Agricultural utilization of sewage sludge. Univ. Minnesota Agric. Exp. Stn. Bull. 606.

Lewis, D.L., Shepard, S., Gattie, D.K., Sanchez, S., and M. Novak. (2000) Enhanced susceptibility to infection from exposure to gases emitted by sewage sludge: a case study. In: *Proceeding of: Biosolids management in the 21st century.* Edited by: H. Kim, O.J. Hao, E.A. Seagren, and A.P. Davis. College Park, MD.

Lindsay, B.J. and T.J. Logan. (1998) Field response of soil physical properties to sewage sludge. *J. Environ. Qual.* **27**: 534-542.

Litvany, W.M. (1998) Compost use in commercial citrus. *Compost News.* **7 (2):** 1-2.

Logan, T.J., and R.L. Chaney. (1983) Utilization of municipal wastewater and sludges on land-metals. In: *Utilization of municipal wastewater and sludge on land.* Edited by: A.L. Page et al. University of California, Riverside, CA.

Logan, T.J., Henry, C.L., Schnoor, J.L., Overcash, M., and D.C. McAvoy. (1999) An assessment of health and environmental risks of trace elements and toxic organics in land-applied municipal solid waste compost. *Compost Science and Utilization.* **7(3)**: 38-53.

Logan, T. 2001. Personal communication.

Loro, P.J., Bergstrom, D.W., and E.G. Beauchamp. (1997) Intensity and duration of denitrification following application of manure and fertilizer to soil. *J. Environ. Qual.* **26:**706-713.

Lowrance, R., Johnson, J.C., Newton, G.L, and R.G. Williams. (1998) Denitrification from soils of a year-round forage production system fertilized with liquid dairy manure. *J. Environ. Qual.* **27:**1504-1511.

Lue-Hing, C., Tata, P., Granato, T.A., Sustich, R., Johnson, R., and R.I. Pietz. (1998) *Sewage Sludge Survey. Association of Metropolitan Sewerage Agencies Members.* Washington, D.C.

MacLeod, J.A., Sanderson, J.B., Douglas, B., and R. Henry. (1998) Use of municipal compost in potato rotation. *National Agricultural Trials, Canadian Composting Council, 1998.*

Madariaga, G.M., and J.S. Angle. (1992) Sludge-born salt effects on survival of *Bradyrhizobium japonicum. J. Environ. Qual.* **21**: 276-280.

Marco, A., Quilchano, C. and A.R. Blaustein. (1999) Sensitivity to nitrate and nitrite in pond breeding amphibians from the Pacific Northwest, USA. *Envtl. Toxicol. & Chem.* **18:**2836-2839.

Mathers, A.C., Stewart, B.A., and J.E. Herring. (1980) Manure and inorganic fertilizer effects on sorghum and sunflower growth on iron-deficient soil. *Agron. J.* **72**:1025-1029.

Maynard, A. (1999) Reducing fertilizer costs with leaf-compost. *BioCycle.* **40(4):** 54-62.

McBride, M.B. (1995) Toxic metal accumulation from agricultural use of sludge: are USEPA regulations protective? *J. Environ. Qual.* **24**: 5-18.

McBride, M.B., Richards, B.K., Steenhuis, T., Russo, J.J, and S. Sauve. (1997) Mobility and solubility of toxic metals and nutrients in soil fifteen years after sludge application *Soil Science.* **162**: 487-500.

McBride, M.B., Richards, B.K., Steenhuis, T.S., G. Spiers. (2000) Molybdenum uptake by forage crops grown on sewage sludge amended soils in the field and greenhouse. *J. Environ. Qual.* **29:** 848-854.

McCaslin, B.D., Davis, J.G., Cihacek, L., and L.A. Schluter (1987) Sorghum yield and soil analysis from sludge-amended calcareous iron-deficient soil. *Agron. J.* **79**: 204-209.

McCoy, S., and B. Cogburn. (2001) Texas makes inroads with highway use of compost. *Biocycle.* **42(2)**: 67-70.

McLeod, R.V. and R.O. Hegg. (1984) Pasture runoff water quality from application of inorganic and organic nitrogen sources. *J. Environ. Qual.* **13:**122-126.

McMurry, S.W., Coyne, M.S., and E. Perfect. (1998) Fecal coliform transport through intact soil blocks amended with poultry manure. *J. Environ. Qual.* **27:** 86-92.

Meckes, M.C., Rice, E.W., Johnson, C.H., and S. Rock. (1995) Assessment of the bacteriological quality of compost from a yard waste processing facility. *Compost Science & Utilization* **3**:6-13.

Midwest Plan Service (1993) *Livestock Waste Facilities Handbook.* Third Edition. Iowa State University. Ames, IA.

Ministry of Agriculture, Fisheries and Food (1998) *Code of Good Agricultural Practice for the Protection of Air.* Welsh Office, Agriculture Department.

Minnesota Environmental Quality Board (1999) *Generic Environmental Impact Statement on Animal Agriculture: a Summary of the Literature related to the Social, Environmental, Economic and Health Effects.* Prepared by the University of Minnesota for the Minnesota Environmental Quality Board.

Moreno, J.L., Hernandez, T., C. Garcia. (1999) Effects of a cadmium-contaminated sewage sludge compost on dynamics of organic matter and microbial activity in an arid soil. *Biology and Fertility of Soils.* **28:** 230-237.

Morgan, H. and D. Simms. (1988) The Shipham report: an investigation into cadmium contamination and its implications for human health. *Science of the Total Environment.* **75**: 1-143.

Mortvedt, J.J. (1985) Plant uptake of heavy metals in zinc fertilizers made from industrial by-products. *J. Environ. Qual.* **14**: 424-427.

Mulla, D.J., Page, A.L., and Ganje, T.J. (1980) Cadmium accumulations and bioavailablity in soils from long-term phosphorus fertilizers. *J. Environ. Qual.* **9:** 408-412.

National Biosolids Partnership (2001) Manual of good practice for biosolids. Interim Final Draft 3-13-01.

National Institute for Occupational Health and Safety (2002). *Guidance for Controlling Potential Risks to Workers Exposed to Class B Biosolids.* DHHS Publication 2002-149. Available at: http://www.cdc.gov/niosh/docs/2002-149/2002-149.html.

National Research Council (1993) *Soil and Water Quality: An Agenda for Agriculture.* National Academy Press. Washington DC.

National Research Council (1996) *Use of Reclaimed Water and Sludge in Food Crop Production.* National Academy Press, Washington, DC.

National Research Council (2000) *Hormonally Active Agents in the Environment.* National Academy Press. Washington, DC.

National Research Council (2002) *Biosolids Applied to the Land: Advancing Standards and Practices.* National Academy Press. Washington, DC.

National Resource Conservation Service (1999) *Agricultural Waste Management Field Handbook.* Soil Conservation Service. 210-AWMFH .

Natural Resources Conservation Service (2000) Conservation Practice Standard: Nutrient Management, Code 590. Available at: http://www.nrcs.usda.gov/technical/ECS/nutrient/590.html.

Ndayegamiye, A., and D. Côté. (1989) Effects of long-term pig slurry and solid cattle manure application on soil chemical and biological properties. *Can. J. Soil Sci.* **69**: 39-47.

New, R. and R. Papworth (1998) Glass contaminant removal from refuse derived compost. Warren Spring Lab., Stevenage (England). Interim Report. April–June 24pp. LR-692 (MR)M; ISBN-0-85624-551-8; NTIS# PB89-197875.

Nichols, D.J., Daniel, T.C., Moore, P.A.,Jr., Edwards, D.R., and D.H. Pote. (1997) Runoff of estrogen hormone 17β-estradiol from poultry litter applied to pasture. *J. Environ. Qual.* **26**: 1002-1006.

Nowak, P., Shepard, R., and F. Madison. (1998) Farmers and manure management: a critical analysis. In: *Animal Waste Utilization: Effective use of Manure as a Soil Resource.* Edited by J.L. Hatfield and B.A.Stewart. Ann Arbor Press. Chelsea, MI.

Oberle, S.L. and D.R. Keeney. (1994) Interactions of sewage sludge with soil-crop-water systems. In: *Sewage Sludge: Land Utilization and the Environment.* Edited by C.E. Clapp, W.E. Larson, and R.H. Dowdy. SSSA Miscellaneous Publication.

O'Connor, G.A. and L.R. McDowell. (1999) *Understanding Fate, Transport, Bioavailablity, and Cycling of Metals in Land-applied Biosolids.* Water Environment Research Foundation. Project 95-REM-3.

O'Connor, D. (2002) *Report of the Walkerton Inquiry: The Events of May 2000 and Related Issues.* Queen's Printer for Ontario. Available at www.walkertoninquiry.com.

O'Handley, R.M., Cockwill, C., McAllister, T.A., Jelinski, M., Morck, D.W., and M.E. Olson. (1999) Duration of naturally acquired giardia and cryptosporidiosis in dairy calves and association with diarrhea. *J. Am. Vet. Med. Assoc.* **214**: 391-396.

O'Leary, M., Rehm, G. and Schmitt, M. (1997). Understanding nitrogen in soils. University of Minnesota Extension Service. FO-3770-GO. Available at http://www.extension.umn.edu/Documents/D/C/DC3770.html.

Oliver, D.P., Hannam, R., Tiller, K.G., Wilhelm, N.S., Merry, R.H., and G.D. Cozens. (1994) The effects of zinc fertilization on cadmium concentration in wheat grain. *J. Environ. Qual.* **23**: 705-711.

Olson, M.E., Guselle, N. ,O'Handley, R., Swift, M.L., McAllister, T.A., Jelinski, M.D., and D.W. Morck. (1997) *Giardia* and *Cryptosporidium* in British Columbia dairy calves. *Can. Vet. J.* **38**: 703-706.

Olson, M.E., Jeanette, G., Phillips, M., Guselle, N., and T.A. McAllister. (1999) Giardia cyst and Cryptosporidium oocyst survival in water, soil, and cattle feces. *J. Environ. Qual.* **28**: 1991-1996.

Ozores-Hampton, M., Schaffer, B., Bryan, H.H., and E.A. Hanlon. (1994) Nutrient concentrations, growth, and yield of tomato and squash in municipal solid waste amended soil. *Hort. Science.* **3:**785-788.

Pain, B.F., and T.H. Misselbrook (1990) Odour and ammonia emissions following the spreading of anaerobically-digested pig slurry on grassland. *Biological Wastes* **34(3):** 259-67.

Parr, J.F. and S.B. Hornick. (1992) Utilization of municipal wastes. In: *Soil Microbial Ecology: Allocations in Agricultural and Environmental Management.* Edited by: F.B. Metting. 45-559. Marcel Dekker, NY.

Pell, A.N. (1997) Manure and Microbes: Public and Animal Health Problem? *J. Dairy Sci.* **80**: 2673-2681.

Peters, J.M and N. T. Basta (1996) Reduction of excessive bioavailable phosphorus in soils by using municipal and industrial wastes. *J. Environ Qual.* **25**:1236-1241.

Peterson, A.E., Speth, P.E., Corey, R.B., Wright, T.H., and P.L Schlecht. (1994) Effect of twelve years of liquid digested sludge application on the soil phosphorous level. In: *Sewage Sludge: Land Utilization and the Environment.* Edited by C.E. Clapp, W.E. Larson, and R.H. Dowdy. SSSA Miscellaneous Publication.

Peterson, E.W., Davis, R.K. and H.A. Orndoff. (2000) 17β-estradiol as an indicator of animal waste contamination in mantled karst aquifers. *J. Environ. Qual.* **29**: 826-834.

Phillips, V.R., B.F. Pain, C.R. Clarkson, and J.V. Klarenbeek (1990) Studies on reducing the odour and ammonia emissions during and after the landspreading of animal slurries. *Farm Buildings and Engineering.* **7**:17-23.

Pillai, S.D., Widmer, K.W., Dowd, S.E., and S.C. Ricke. (1996) Occurrence of airborne bacteria and pathogen indicators during land application of sewage sludge. *Applied Environmental Microbiology*. **62**: 296-299.

Puckett, L.J. (1994) Nonpoint and point sources of nitrogen in major watersheds of the United States. National Water Quality Assessment. *U.S. Geological Survey*. Water-Resources Investigations Report No. 94-4001.

Raven, K.P. and R.H. Loeppert. (1997) Trace element composition of fertilizers and soil amendments. *J. Environ. Qual.* **26**: 551-557.

Reeves, S., Thomberry, F., and J.B. Carey. (1998) Poultry Waste Management Handbook. Texas Poultry Federation and Texas Agricultural Extension Service Texas A&M University.

Richards, B.K., Steenhuis, T.S., Peverly, J.H., M.B. McBride. (1998) Metal mobility at an old, heavily loaded sludge application site. *Environmental Pollution.* **99:** 365-377.

Riggle, D. (1996) Compost teas in agriculture. *BioCycle.* **37(12):** 65-67.

Ritter, W.F. (2000) Potential impact of land application of by-products on ground water and surface water quality. In: *SSSA Book Series: 6: Land Application of Agricultural, Industrial, and Municipal By-Products.* Edited by J.F. Power and W.A. Dick.

Rodd, V., Warman, P., Hicklenton, P., Webb, K., Halsey, R., Friesen, B., Breau, P., Antler, S., and Hennessey, T. (1998) Comparison of source-separated municipal solid waste compost and solid manure with and without fertilizer N. *National Agricultural Trials, Canadian Composting Council, 1998.*

Rouse, J.D., Bishop, C.A., and J. Struger. (1999) Nitrogen pollution: an assessment of its threat to amphibian survival. *Environ. Health Perspect.* **107:**799-803.

Rudolfs, W., Falk, L.L., and R.A. Ragotzkie. (1950) Literature review of the occurrence and survival of enteric pathogenic, and relative organisms in soil, water, sewage, and sludge and on vegetation. *Sewage and Industrial Wastes*. **22** (10): 1261-1281.

Rynk, R. (1992) *On-Farm Composting Handbook.* Northeast Regional Agricultural Engineering Service. NRAES-54. Ithaca, NY.

SB1217 Interagency Group. (2001). Guidance Document. Available at www.soil.ncsu.edu/interagency/guidance/chapter1.html.

Scharp, M. (2000) Why Colorado farmers want New York City biosolids. Biocycle. 41(7): 71-76.

Schiffman, S.S., Walker, J.M., Dalton, P., Lorig, T.S., Raymer, J.H., Shusterman, D., and M. Williams. (2000) Potential health effects of odor from animal operations, wastewater treatment, and recycling of byproducts. *J. of Agromedicine.* 7(1): 7-80.

Schjønning, P., Christensen, B.T., and B. Cartensen. (1994) Physical and chemical properties of sandy loam receiving animal manure, mineral fertilizer of no fertilizer for 90 years. *European J. Soil Sci.* 45: 257-268.

Sharpley, A, Meisinger, J.J., Breeuwsma, A., Sims, J.T., Daniel, T.C., and J.S. Schepers. (1994) Impacts of animal manure management on ground and surface water quality. In: *Animal Waste Utilization: Effective use of Manure as a Soil Resource.* Edited by J.L. Hatfield and B.A.Stewart. Ann Arbor Press. Chelsea, MI.

Sharpley, A.N. and I. Sisak. (1997) Differential availability of manure and organic sources of phosphorus in soil. *Soil Sci Soc. Am. J.* **61**: 1503-1508.

Sharpley, A., Daniel, T., Sims, T., Lemunyon, J., Stevens, R., and Parry, R. (1999) *Agricultural Phosphorus and Eutrophication.* United States Department of Agriculture. ARS-149.

Shelton, J.E. (1991) Using municipal solid waste compost. AG-439-19. North Carolina Cooperative Extension Service.

Shiralipour, A., McConnell, D.B., W. Smith. (1992) Uses and benefits of MSW compost: A review and an assessment. *Biomass and Bioenergy.* **3(3&4)**: 267-279.

Shiralipour, A., Faber, B., and M. Chrowstowski. (1996) Greenhouse broccoli and lettuce growth using cocomposted biosolids. *Compost Science and Utilization.* **4(3):** 38-43.

Shore, L.S., Correll, D.L., and P.K. Chakraborty. (1995) Relationship of fertilization with chicken manure and concentrations of estrogens in small streams. In: *Animal Waste and the Land-Water Interface.* Edited by Kenneth Steele. CRC Press. Boca Raton, FL.

Shore, L.S., Harel-Markowitz, E., Gurevich, M. and M. Shemesh. (1993) Factors affecting the concentration of testosterone in poultry litter. *J Environ. Sci. Health*, **A28(8):**1737-1749.

Sims, J.T. and D.C. Wolf (1994) Poultry waste management: agricultural and environmental issues. *Advances in Agronomy.* **52**:51-59.

Sims, J.T. (1995) Characteristics of animal wastes and waste-amended soils: an overview of the agricultural and environmental issues. In: *Animal Waste and the Land-Water Interface*, edited by Kenneth Steele. CRC Press. Boca Raton, FL.

Sims, J.T., Dentel, S.K., Coale, F.J., Maguire, R.O., and Y. Khadar. (1999) *Assessing the availability and environmental impact of phosphorus in soils amended with biosolids.* Submitted to: Metropolitan Washington Council of Governments.

Sloan, J.J, Dowdy, R.H., and M.S. Dolan. (1998). Recovery of biosolids-applied metals sixteen years after application. *J. Environ Qual.* 27:1312-1317.

Smith, D.R. (1996) *Agricultural recycling of sewage sludge and the environment.* Wallingford, United Kingdom: CAB International.

Smith, D.R., Moore, P.A., Griffis, C.L., Daniel, T.C., Edwards, D.R., and Boothe, D.L. (2001) Effects of alum and aluminum chloride on phosphorus runoff from swine manure. *J. Environ Qual.* **30:** 992-998.

Soares, H.M., Cardenas, B., Weir, D. and M.S. Switzenbaum. (1995) Evaluating pathogen regrowth in biosolids compost. *BioCycle.* **36(6):** 70-75.

Sommerfeldt, T.G., Chang, C., and T. Entz. (1988) Long-term annual manure applications increase soil organic matter and nitrogen, and decrease carbon to nitrogen ratio. *Soil Sci. Soc. Am. J.* **52**:1668-1672.

Sorber, C.A., and B.E. Moore (1986) Survival and transport of pathogens in sludge-amended soil, a critical literature review. Report No. EPA/600/2-87/028. USEPA. Cincinnati, OH.

Sort, X. and J.M. Alcaniz. (1996) Contribution of sewage sludge to erosion control in the rehabilitation of limestone quarries. *Land Degradation and Development* **7:**69-76.

Spallacci, P., and V. Boshi. (1985) Long-term effects of the land spreading of pig and cattle slurries on the accumulation and availability of nutrients. In *Long-term Effects of Sewage Sludge and Farm Slurries Applications.* Edited by J. H. Williams, G. Guidi, and P. L'hermite. Elsevier Applied Sci. Pub., NY.

Stehouwer, R. (2001) Meeting minutes from the annual meeting of NE 1001. University of New Hampshire. Durham, New Hampshire. Available at: http://www.cfe.cornell.edu/wmi/Sludge/NEreport.pdf

Stephens, J.M. and S.R. Kostewicz. (1994) Response of cucumber to organic soil amendments. *Proc. Fla. State Hort. Soc.* **107**:382-384.

Stewart, B.A., Robinson, C.A., and D.B. Parker. (2000) Examples and case studies of beneficial reuse of beef cattle by-products. In: *SSSA Book Series: 6: Land Application of Agricultural, Industrial, and Municipal By-Products*. Edited by J.F. Power and W.A. Dick.

Stoddard, C.S., Coyne, M.S. and J.H. Grove. (1998) Fecal bacteria survival and infiltration through a shallow agricultural soil: timing and tillage effects. *J. Environ. Qual.* **27**:1516-1523.

Stowe, D.W. and C. Caudill. (1994) Long term biosolids research on dryland wheat farms. Proceedings of the Water Environment Federation: 67th Annual Conference and Exposition. Chicago, IL.

Strauch, D. (1991) Survival of pathogenic micro-organism and parasites in excreta, manure and sewage sludge. *Rev. Sci. Tech. Off. Int. Epizoot.* **10**: 813-846.

Sullivan, D. (1998) Fertilizing with biosolids. PNW-508. Available at: http://eesc.orst.edu/agcommwebfile/edmat/PNW508.pdf

Susarla, S., Conllette, T.W., Garrison, A.W., Wolfe, N.L., and S.C. McCutcheon. (1999) Perchlorate identification in fertilizers. *Environmental Science and Technology.* **33**: 3469-3472.

Tester, C.F. (1990) Organic amendment effects on physical and chemical properties of a sandy soil. *Soil Sci. Soc. Am. J.* **54**: 827-831.

Texas Agricultural Extension Service (2001) *Nutrient Management Planning Certification 2001.*

Topp, E. and A. Starratt (2000) Rapid mineralization of endocrine-disrupting chemical 4-nonylphenol in soil. *Environmental Toxicology and Chemistry.* **19:**313-9.

Turner, M.S., Clark, G.A., Stanley, C.D., and A.G. Smajstrla. (1994) Physical characteristics of a sandy soil amended with MSW compost. *Soil And Crop Sci. Soc. Florida Proc.* **53:** 24-26.

U.S. Department of Agriculture (1997) *Agricultural Resources and Environmental Indicators, 1996-97.* Economic Research Service. Agricultural Handbook No. 712.

U.S. Department of Agriculture (1998) *Agricultural Uses of Municipal, Animal, and Industrial Byproducts.* Agricultural Research Service. Conservation Report No. 44.

U.S. Department of Agriculture/Environmental Protection Agency (1999) *Unified National AFO Strategy Executive Summary.* Available at: www.epa.gov/owm/finafost.htm.

U.S. Environmental Protection Agency (1977) *Effect of Land Disposal Applications of Municipal Wastes on Crop Yields and Heavy Metal Uptake.* EPA –600/2-77-014.

U.S. Environmental Protection Agency (1990a) *National Survey of Pesticides in Drinking Water Wells: Phase 1 Report.* EPA 570/9-90-015.

U.S. Environmental Protection Agency (1990b). *National Sewage Sludge Survey: Availability of Information and Data, and Anticipated Impacts in Proposed Regulations.* Proposal Rule 40 CFR 503. Federal Register **55 (218)**: 47210-47283, Nov. 9, 1990.

U.S. Environmental Protection Agency (1993) *Guidance for Writing Permits for the Use or Disposal of Sewage Sludge (Draft)* Washington, DC.

U.S. Environmental Protection Agency (1995) *A Guide to the Biosolids Risk Assessments for the EPA Part 503 Rule.* EPA832-B-93-005. Washington DC.

U.S. Environmental Protection Agency (1997a) *Municipal Solid Waste Factbook.* Available at: http://www.epa.gov/epaoswer/non-hw/muncpl/factbook/internet/.

U.S. Environmental Protection Agency (1997b) *Special Report on Endocrine Disruptors.* Washington, DC.

U.S. Environmental Protection Agency (1998) *Environmental Impacts of Animal Feeding Operations.* Washington, DC.

U.S. Environmental Protection Agency (1999a) *Biosolids Generation, Use, and Disposal in the United States.* EPA530-R-99-009. Washington, DC.

U.S. Environmental Protection Agency (1999b) *Standards for the Use or Disposal of Sewage Sludge: Proposed Rule.* Federal Register **64 (246)**: 72045-72062.

U.S. Environmental Protection Agency (1999c) *Estimating Risk from Contaminants Contained in Agricultural Fertilizers, Draft Report.*

U.S. Environmental Protection Agency (2000a). *Guide to Field Storage of Biosolids.* EPA/832-B-00-007. Washington DC.

U.S. Environmental Protection Agency (2000b) *Requirements for Zinc Fertilizers Made From Recycled Hazardous Secondary Material; Proposed Rule.* Federal Register **65(229):** 70953-70982, November 28, 2000.

U.S. Environmental Protection Agency (2000c) *Water Quality Conditions in the United States.* EPA841-F-00-006. Washington DC.

U.S. Environmental Protection Agency (2001a) *National Pollutant Discharge Elimination System Permit Regulations and Effluent Limitations Guidelines and Standards for Concentrated*

Animal Feeding Operations; Proposed Rule. Federal Register. **66(9):** 2959-3145, January 12, 2001.

U.S. Environmental Protection Agency (2001b) *Perchlorate*. Available at: http://www.epa.gov/ogwdw000/ccl/perchlor/perchlo.html.

U.S Environmental Protection Agency (2001c). *Survey of Fertilizers and Related Materials for Perchlorate (ClO_4^-).* EPA/600/R-01/047. Washington DC.

U.S. Environmental Protection Agency (2001d). *Environmental Assessment of Proposed Revisions to the National Pollutant Discharge Elimination System Regulations and the Effluent Guidelines for Concentrated Animal Feeding Operations.* EPA-821-B-01-001. Washington DC.

Valcour, J.E., Michel, P., McEwen, S.A., and J.B. Wilson. (2002) Associations between indicators of livestock farming intensity and incidence of human Shiga toxin-producing *Escherichia coli. Emerging Infections Diseases.* Vol 18: No 3. Available at: www.cdc.gov/ncidod/EID/vol18no3/01-0159.htm.

Van der Voet, E., van Egmond, L., Kleijn, R., and G. Huppes. (1994) Cadmium in the european community - a policy oriented analysis. *Waste Management and Research* **12:** 507-526.

Van der Watt, H.H., Sumner, M.E., and M.L. Cabrera. (1994) Bioavailability of copper, manganese, and zinc in poultry litter. *J. Environ. Qual.* **23**:43-49.

Veenhuizen, M.A., Eckert, D.J., Elder, K., Johnson, J., Lyon, W.F., Mancl, K.M., and G. Schnitkey. (1992) Ohio livestock manure and wastewater management guide. (Bulletin 604-92) Ohio State University, OH.

Wall, G.J., Grant, B.A., D.J. King, and N. McLaughlin (1997) *The Effects of Livestock Manure Application Methods on Water Quality, Focusing on Nitrogen and Bacteria Transport in Soil.* Final Report RES/MAN-005-97. Agriculture and Agri-Food Canada.

Washington State Department of Ecology. (1997) *Screening Survey for Metals in Fertilizers and Industrial by-Product Fertilizers in Washington State*. Ecology Publication No. 97-341. Olympia, WA.

Washington State Department of Ecology (1999) *Screening Survey for Metals and Dioxins in Fertilizer Products and Soils in Washington State.* Ecology Publication No. 99-309. Olympia, WA.

Washington State Department of Ecology (2000) *Biosolids Management Guidelines for Washington State*. Publication No. 93-80 (revised July 2000). Available at: http://www.puyallup.wsu.edu/soilmgmt/BMG.htm.

Water Environment Association of Ontario. (2001). *Fate and Significance of Selected Contaminants in Sewage Biosolids Applied to Agricultural Land Through Literature Review and Consultation with Stakeholder Groups.* Available at: www.weao.org.

WaterUK. (2000) Sludge to agricultural land: technical notes. Prepared by the recycling network at WaterUK.

Weinberg Group (1998a) *Industry and Literature Survey of Nutritive and Non-Nutritive Elements in Inorganic Fertilizer Materials*. Prepared for the Fertilizer Institute, Washington D.C.

Weinberg Group (1999) *Perchlorate Information Summaries.* Prepared for: The Fertilizer Institute.

Weinberg Group (2000) *Health Risk Evaluation of Select Metals in Inorganic Fertilizers Post Application.* Prepared for the Fertilizer Institute.

Withers, P.J.A., Clay, S.A., and V.G. Breeze. (2001). Phosphorus transfer following application of fertilizer, manure and sewage sludge. *J. Environ. Qual.* **30**:180-188.

Wood, C.W., and J.A. Hattey. (1995) Impacts of Long-Term Manure Applications on Soil Chemical, Microbiological, and Physical Properties. In: *Animal Waste and the Land-Water Interface*. Edited by Kenneth Steele. CRC Press. Boca Raton, FL.

Woodbury, P.B. (1992) Trace elements in municipal solid waste compost: a review of potential detrimental effects on plants, soil biota, and water quality. *Biomass and Bioenergy* **3(3-4)**:239-259.

Yamakawa. 2000. Personal communication.

Yanko, W.A. (1988) *Occurrence of Pathogens in Distribution and Marketing Municipal Sludges*. EPA/600/1-87/014. Washington, DC.

Zebarth, B.J., Neilsen G.H., Hogue, E., and D. Neilsen. (1999) Influence of organic waste amendments on selected soil physical and chemical properties. *Canadian Journal of Soil Science.* **79**: 501-504.

Zhao, T., Doyle, M.P., Shere, J, and L. Garber. (1995) Prevalence of enterohemorrhagic Escherichia coli O157:H7 in a survey of dairy herds. *Applied and Environmental Microbiology*. **61**:1290-1293.

Zupanci K.S., Hattey, J.A., Mikkelsen, R., Zhang, H.A., Hamilton, D., and M. Kizer. (1998) Determination of ammonia flux from swine effluent applications in the southern great plains. In: *Procs. of the International Conference on Odor, Water Quality, Nutirent Managemnt and Socioeconomic Issues.* Des Moines, IA. **2**: 565-569.

WERF Subscribers

The following agencies and corporations are WERF's charter stormwater research supporters:

Stormwater Agency

California
Monterey, City of
Santa Rosa, City of

Colorado
Boulder, City of

Georgia
Griffin, City of

Kansas
Overland Park, City of

Kentucky
Louisville & Jefferson County Metropolitan Sewer District

North Carolina
Cary, Town of
Charlotte, City of/Mecklenburg County

Pennsylvania
Philadelphia, City of

Corporate

BaySaver Inc.
Vortechnics Inc.
WWETCO, LLC

State

Fresno Metropolitan Flood Control District, Calif.
Urban Drainage & Flood Control District, Colo.

Wastewater Utility

Alabama
Montgomery Water Works & Sanitary Sewer Board

Alaska
Anchorage Water & Wastewater Utility

Arizona
Gila Resources
Glendale, City of, Utilities Department
Mesa, City of
Phoenix Water Services Department
Pima County Wastewater Management

Arkansas
Little Rock Wastewater Utility

California
Central Contra Costa Sanitary District
Contra Costa Water District
Crestline Sanitation District
Delta Diablo Sanitation District
Dublin San Ramon Services District
East Bay Municipal Utility District
El Dorado Irrigation District
Escondido, City of
Fairfield-Suisun Sewer District
Irvine Ranch Water District
Las Virgenes Municipal Water District
Lodi, City of
Los Angeles, City of
Los Angeles County, Sanitation Districts of
Orange County Sanitation District
Palo Alto, City of
Riverside, City of
Sacramento Regional County Sanitation District
San Diego Metropolitan Wastewater Department, City of
San Francisco, City & County of
San Jose, City of
Santa Barbara, City of
Santa Rosa, City of
South Bayside System Authority
South Orange County Wastewater Authority
Union Sanitary District

Colorado
Boulder, City of
Colorado Springs, City of
Littleton/Englewood Water Pollution Control Plant
Metro Wastewater Reclamation District, Denver

Connecticut
New Haven, City of, WPCA

District of Columbia
District of Columbia Water & Sewer Authority

Florida
Broward, County of
Fort Lauderdale, City of
Gainesville Regional Utilities
JEA
Kissimmee, City of, Department of Water Resources
Miami-Dade Water & Sewer Authority
Orange County Utilities Department
Orlando, City of
Reedy Creek Improvement District
Seminole County Environmental Services
St. Petersburg, City of
Stuart Public Utilities
Tallahassee, City of
Tampa, City of
West Palm Beach, City of

Georgia
Atlanta Wastewater Services
Augusta, City of
Clayton County Water Authority
Cobb County Water System
Columbus Water Works
Fulton County
Gwinnett County Department of Public Utilities
Macon Water Authority
Savannah, City of

Hawaii
Honolulu, City and County of

Illinois
American Bottoms Wastewater Treatment Plant
Dupage, County of
Greater Peoria Sanitary District
Metropolitan Water Reclamation District of Greater Chicago
Wheaton Sanitary District

Indiana
Fort Wayne, City of
Indianapolis, City of

Iowa
Ames, City of
Des Moines Metro Wastewater Reclamation Authority

Kansas
Johnson County Unified Wastewater Districts
Unified Government of Wyandotte County/Kansas City, City of

Kentucky
Louisville & Jefferson County Metropolitan Sewer District

Louisiana
Sewerage & Water Board of New Orleans

Maine
Bangor, City of

Maryland
Anne Arundel County Bureau of Utility Operations
Howard County Department of Public Works
Washington Suburban Sanitary Commission

Massachusetts
Boston Water & Sewer Commission
Massachusetts Water Resources Authority
Upper Blackstone Water Pollution Abatement District

Michigan
Ann Arbor, City of
Detroit, City of

Grand Rapids, City of

Holland Board of Public Works

Lansing, City of

Owosso Mid-Shiawassee County WWTP

Saginaw, City of

Wayne County Department of Environment

Wyoming, City of

Missouri

Independence, City of

Kansas City Missouri Water Services Department

Little Blue Valley Sewer District

Metropolitan St. Louis Sewer District

Nebraska

Lincoln Wastewater System

Nevada

Henderson, City of

New Jersey

Passaic Valley Sewerage Commissioners

New York

New York City Department of Environmental Protection

North Carolina

Cary, Town of

Charlotte/Mecklenburg Utility Department

Durham, City of

Metropolitan Sewerage District of Buncombe County

Orange Water & Sewer Authority

Ohio

Akron, City of

Butler County Department of Environmental Services

Columbus, City of

Metropolitan Sewer District of Greater Cincinnati

Northeast Ohio Regional Sewer District

Oklahoma

Tulsa, City of

Oregon

Clean Water Services

Eugene/Springfield Water Pollution Control

Water Environment Services

Pennsylvania

Philadelphia, City of

University Area Joint Authority, State College

South Carolina

Charleston Commissioners of Public Works

Mount Pleasant Waterworks & Sewer Commission

Spartanburg Sanitary Sewer District

Tennessee

Cleveland, City of

Knoxville Utilities Board

Murfreesboro Water & Sewer Department

Nashville Metro Water Services

Texas

Amarillo, City of

Austin, City of

Dallas Water Utilities

Denton, City of

El Paso Water Utilities

Fort Worth, City of

Gulf Coast Waste Disposal Authority

Houston, City of

San Antonio Water System

Trinity River Authority

Utah

Salt Lake City Corporation

Virginia

Alexandria Sanitation Authority

Arlington, County of

Fairfax County Virginia

Hampton Roads Sanitation District

Henrico, County of

Hopewell Regional Wastewater Treatment Facility

Loudoun County Sanitation Authority

Prince William County Service Authority

Richmond, City of

Washington

Edmonds, City of

Everett, City of

King County Department of Natural Resources

Seattle Public Utilities

Wisconsin

Green Bay Metro Sewerage District

Madison Metropolitan Sewerage District

Milwaukee Metropolitan Sewerage District

Racine, City of

Sheboygan Regional Wastewater Treatment

Wausau Water Works

Australia

Sydney Water Corporation

Canada

Toronto, City of, Ontario

Mexico

Servicios de Agua y Drenaje de Monterrey, I.P.D.

United Kingdom

Yorkshire Water Services Limited

Corporate

The ADVENT Group Inc.

Alan Plummer & Associates

Alpine Technology Inc.

Aquateam–Norwegian Water Technology Centre A/S

Black & Veatch

Boyle Engineering Corporation

BPR CSO

Brown & Caldwell

Burns & McDonnell

The Cadmus Group

Camp Dresser & McKee Inc.

Carollo Engineers Inc.

Carpenter Environmental Associates Inc.

CDS Technologies Inc.

Chemtrac Systems Inc.

CH2M HILL

Clancy Environmental Consultants Inc.

Damon S. Williams Associates, LLC

Earth Tech Inc.

Ecolab Water Care Services

EMA Inc.

The Eshelman Company Inc.

Finkbeiner, Pettis, & Strout Inc.

Frontier Geosciences Inc.

ftn Associates Inc.

Gannett Fleming Inc.

GE Betz

Golder Associates Inc.

Greeley and Hansen LLC

The HACH Company

Hazen & Sawyer, P.C.

HDR Engineering Inc.

HNTB Corporation

HydroQual Inc.

Insituform Technologies Inc.

Institute for Environmental Technology & Industry, Korea

Jacobson Helgoth Consultants Inc.

Jason Consultants Inc.

Jordan, Jones, & Goulding Inc.

KCI Technologies Inc.

Kelly & Weaver, P.C.

Kennedy/Jenks Consultants

Komline Sanderson Engineering Corporation

Lawler, Matusky & Skelly Engineers, LLP

Limno-Tech Inc.

Lombardo Associates Inc.

Malcolm Pirnie Inc.

McKim & Creed

MWH

Odor & Corrosion Technology Consultants Inc. (OCTC)

ONDEO Degremont Inc.

PA Government Services Inc.

Parametrix Inc.

Parsons

Post, Buckley, Schuh & Jernigan

The RETEC Group

R.M. Towill Corporation

Royce Technologies

Sear-Brown

SYNAGRO

Tetra Tech Inc.

Trojan Technologies Inc.

URS Corporation

USFilter NATC

Wade-Trim Inc.

Weston Solutions Inc.

Woodard & Curran

Woodruff & Howe Environmental Engineering Inc. (WHEE)

WRc/D&B, LLC

Industry

American Electric Power

BP-Amoco Corporation

Chevron Research & Technology Company

The Coca-Cola Company

Dow Chemical Company

DuPont Company

Eastman Chemical Company

Eastman Kodak Company

Merck & Company Inc.

ONDEO Services

Procter & Gamble Company

Reliant Energy

Severn Trent Services Inc.

Shell Global Solutions

Thames Water Plc

United Water Services LLC

Board of Directors

Research Council

Stormwater Technical Advisory Committee

WERF Product Order Form

As a benefit of joining the Water Environment Research Foundation, subscribers are entitled to receive one complimentary copy of all final reports and other products. Additional copies are available at cost (usually $10). To order your complimentary copy of a report, please write "free" in the unit price column. WERF keeps track of all orders. If the charge differs from what is shown here, we will call to confirm the total before processing.

Name ______________________ Title ______________________

Organization ______________________

Address ______________________

City __________ State __________ Zip Code __________ Country __________

Phone __________ Fax __________ Email __________

Stock #	Product	Quantity	Unit Price	Total
			Postage & Handling	
			VA Residents Add 4.5% Sales Tax	
			Canadian Residents Add 7% GST	
			TOTAL	

Method of Payment:

(All orders must be prepaid.)

❑ Check or Money Order Enclosed

❑ Visa ❑ Mastercard ❑ American Express

Account No. ______________________ Exp. Date __________

Signature ______________________

Shipping & Handling:

Amount of Order	United States	Canada & Mexico	All Others
Up to but not more than:	Add:	Add:	Add:
$20.00	$5.00*	$8.00	50% of amount
30.00	5.50	8.00	40% of amount
40.00	6.00	8.00	
50.00	6.50	14.00	
60.00	7.00	14.00	
80.00	8.00	14.00	
100.00	10.00	21.00	
150.00	12.50	28.00	
200.00	15.00	35.00	
More than $200.00	Add 20% of order	Add 20% of order	

*minimum amount for all orders

Note: Please make checks payable to the Water Environment Research Foundation.

To Order (Subscribers Only):

Log on to www.werf.org and click on the "Product Catalog."

Phone: (703) 684-2470
Fax: (703) 299-0742.

WERF
Attn: Subscriber Services
601 Wythe Street
Alexandria, VA 22314-1994

To Order (Non-Subscribers):

Non-subscribers may be able to order WERF publications either through WEF (www.wef.org) or IWAP (www.iwap.co.uk). Visit WERF's website at www.werf.org for details.

IWA Publishing's authorised EU representative for General Product Safety Regulations is Diane D'Arras, 15 rue Duret, 75116 Paris, France, e-mail: safety@iwap.co.uk.

Printed and bound by CPI Group (UK) Ltd, Croydon, CR0 4YY
12/05/2026
02108865-0015